Lotus Illustrated DICTIONARY of BIO TECHNOLOGY

Complied and Edited by:
Niel Sanders

Lotus PRESS
4735/22, Prakash Deep Building
Ansari Road, Daryaganj,
New Delhi - 110002

Lotus *Illustrated*

DICTIONARY *of* BIO TECHNOLOGY

ISBN 81 89093 16 9

Published by:

Lotus Press Publishers & Distributors
Unit No. 220, 2nd Floor, 4735/22,
Prakash Deep Building, Ansari Road, Darya Ganj,
New Delhi- 110002, Ph.: 32903912, 23280047, 09811838000
• E-mail : lotuspress1984@gmail.com www.lotuspress.co.in

Printed at : Lotus Press Publishers & Distributors, New Delhi

PREFACE

Biotechnology is the use of whole or parts of organisms, such as cells, genetic materials or enzymes, to make or modify products, to improve valuable traits in plants or animals, or to develop organisms for specific applications. It also encompasses research and development of natural products from organisms for agricultural, biomedical and industrials uses. Biotechnology is a very wide field that offers particularly promising opportunities for both developing and developed countries. It requires the co-operation of multinational scientific communities and draws strength from various disciplines of science, including environmental biotechnology, marine biology to cell and molecular biology.

The term biotechnology has become particularly familiar after the advent of genetic engineering during the 1970s. Most of the 'new' biotechnology uses organisms that have been genetically altered to enable them to work more effectively than earlier, or to function in entirely new ways.

The dictionary is an attempt to enhance the reader's knowledge in this comparatively new discipline. The endeavour behind collating such a huge repertoire of terms related to the subject is to provide the readers with an easy and handy referral. The pictures and examples used in the dictionary enhance the explanation and enable the user to cull out the best and the highest level of understanding. The language used is deliberately and strictly kept to be simple, so as to cater to both the beginners as well as the advanced level of learners. The knowledge gained from the dictionary can act as a base for the readers who are quite new to the subject and thus make them well equipped with appropriate level of knowledge to venture into the subject.

■ **2mm plasmid**

a naturally occurring, double-stranded, circular DNA plasmid (6 318 bp) found in the nuclei of yeast. Many yeast plasmid vectors are derived from the 2mm plasmid. a.k.a. 2mm circle; 2 m-plasmid; 2-micron plasmid.

■ **3´-extension**

a short single-stranded nucleotide sequence on the 3´-hydroxyl end of a double-stranded DNA molecule. a.k.a. 3´ protruding end; 3´ sticky end; 3´ overhang.

■ **5´ end**

the phosphate group that is attached to the 5´ carbon atom of a sugar (ribose or deoxyribose) of the terminal nucleotide of a nucleic acid molecule.

■ **5´-extension**

a short single-stranded nucleotide sequence on the 5´-hydroxyl end of a double-stranded DNA molecule. a.k.a. 5´ protruding end; 5´ sticky end; 5´ overhang.

■ **A**

Adenine residue, in either DNA or RNA.

■ **abiotic stress**

the effect of non-living factors which can harm living organisms. These non-living factors include drought, extreme temperatures, pollutants, etc.

■ **abscisic acid**

a plant growth regulator involved in abscission, dormancy, stomatal opening/closure, and inhibition of seed germination. It also affects the regulation of somatic cell embryogenesis in some plant species.

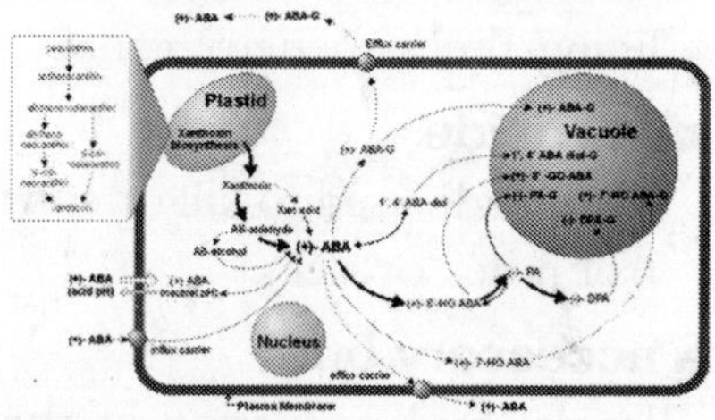

■ **absciss**

the horizontal axis of a graph. *cf* ordinate.

■ **absorb**

to suck up, or to take in. In the cell, materials are taken in (absorbed) from a solution. *cf* adsorb.

■ **absorption**

in general: the process of absorbing; taking up of water

and nutrients by assimilation or imbibition. The taking up by capillary, osmotic, chemical or solvent action, such as the taking up of a gas by a solid or liquid, or taking up of a liquid by a solid. *cf* adsorption.In biology: the movement of a fluid or a dissolved substance across a cell membrane. In plants: water and mineral salts are absorbed from the soil by roots. In animals: solubulized food material is absorbed into the circulatory system through cells lining the alimentary canal.

■ acaricide

a pesticide used to kill or control mites or ticks.

■ accessory bud

lateral bud occurring at the base of a terminal bud or at the side of an auxiliary bud.

■ acclimatisation

the adaptation of a living organism (plant, animal or micro-organism) to a changed environment that subjects it to physiological stress. Acclimatisation should not be confused with adaptation (q.v.). *cf* acquired.

■ acellular

describing tissues or organisms that are not made up of separate cells but often have more than one nucleus. *cf* syncytium.

■ acentric chromosome

chromosome fragment lacking a centromere.

■ acetyl co-enzyme A

a compound formed in the mitochondria when an acetyl group (CH_3CO-) - derived from breakdown of fats, proteins, or carbohydrates - combines with the thiol group (-SH) of co-enzyme A.

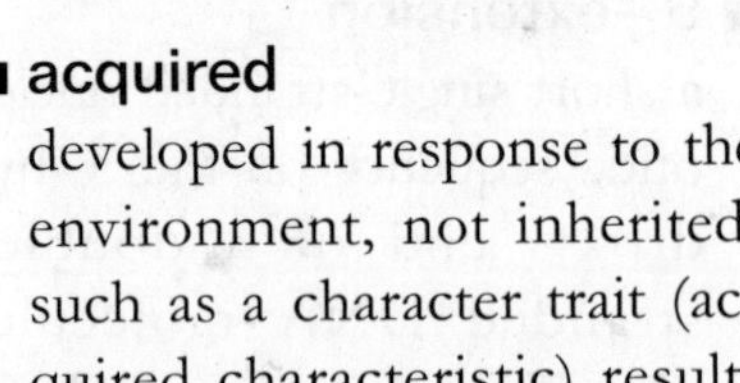

■ acquired

developed in response to the environment, not inherited. such as a character trait (acquired characteristic) resulting from environmental effect(s). *cf* acclimatisation.

■ acridine dyes

a class of positively charged polycyclic molecules that intercalate into DNA and induce frameshift mutations.

■ acrocentric

a chromosome that has its centromere near the end.

■ acropetal

1. developing or blooming in succession towards the apex, such as leaves or flowers developing acropetally.

2. transport or movement of substances towards the apex, such as the movement of water through the plant. The opposite tendency is termed basipetal.

■ acrosome

an apical organelle in the head of a spermatozoon, q.v.

■ actin

one of the two contractile proteins in muscle (the other being myosin). Actin is also found in the microfilaments that form part of the cytoskeleton of all cells.

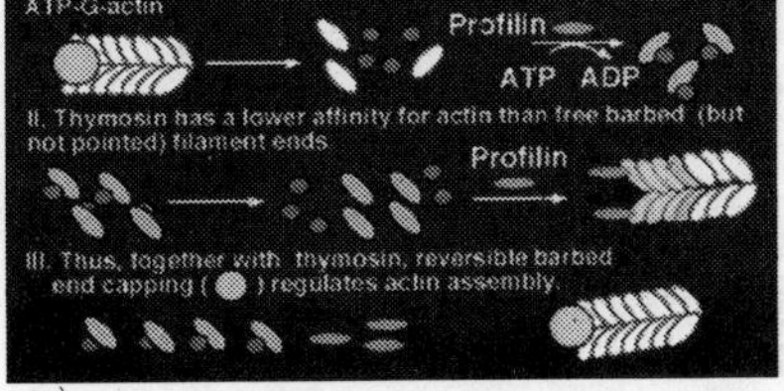

■ activated charcoal

charcoal which has been treated to remove hydrocarbons and to increase its adsorptive properties. It acts by condensing and holding a gas or solute onto its surface; thus inhibitory substances in nutrient medium may be adsorbed to charcoal included in the medium. Rooting factors such as phenolamines present as contaminants in charcoal may stimulate growth *in vitro*. Its addition to rooting medium may stimulate root initiation in some plant species. Activated charcoal may differ in origin and in composition. *cf* charcoal; phenolic oxidation.

■ activator

1. a substance or physical agent that stimulates transcription of a specific gene or operon.

2. a compound that, by binding to an allosteric site on an enzyme, enables the active site of the enzyme to bind to the substrate.

■ active collection

in PGR: Defined in the International Undertaking on Plant

Genetic Resources (FAO, 1983) as a collection which complements a base collection (q.v.) and is a collection from which seed samples are drawn for distribution, exchange and other purposes such as multiplication and evaluation.

■ **active site**

1. a site on the surface of a catalyst at which activity occurs.
2. the site on the surface of an enzyme molecule that binds the substrate molecule.

■ **adaptation**

adjustment of a population to changed environment over generations, associated (at least in part) with genetic changes resulting from selection imposed by the changed environment. *Not* acclimatisation, q.v.

■ **adaptation traits**

the complex of traits related to reproduction and survival of the individual in a particular production environment. Adaptation traits contribute to individual fitness; they are the traits subjected to selection during the evolution of animal genetic resources. By definition, these traits are also important to the ability of the animal genetic resource to be sustained in the production environment.

■ **adaptive radiation**

the evolution of new forms, sub-species or species from one species of plant or animal in order to exploit new habitats or food sources. a.k.a. divergent evolution.

■ **adaptor**

1. a synthetic double-stranded oligonucleotide that has a blunt end, while the other end has a nucleotide extension that can base pair with a cohesive end created by cleavage of a DNA molecule with a specific type II restriction endonuclease. The blunt end of the adaptor can be ligated to the ends of a target DNA molecule and the construct can be cloned into a vector by using the cohesive ends of the adaptor.
2. a synthetic single-stranded oligonucleotide that, after self-hybridisation, produces a molecule with cohesive ends and an internal restriction endonuclease site. When the adaptor is inserted into a clon-

ing vector by means of the cohesive ends, the internal sequence provides a new restriction endonuclease site.

■ **addendum**

in formulation of tissue culture media: an item or a constituent substance to be added.

■ **additive allelic effects**

effects of alleles at a locus, where the heterozygote is exactly intermediate between the two homozygotes.

■ **additive gene effects**

additive allelic effects summed across all the loci that contribute to genetic variation in a quantitative trait.

■ **adenine**

a white crystalline purine base. A constituent of DNA and RNA and nucleotides such as ADP and ATP. A B-group vitamin (B_4) generally available as $C_5H_5N_5.3H_2O$, m.p. 189.13. It is added to some tissue culture media, as adenine sulphate, to promote shoot formation and for its weak cytokinin effect. It is present in plant tissues combined with aminoamide, phosphoric acids and D-ribose.

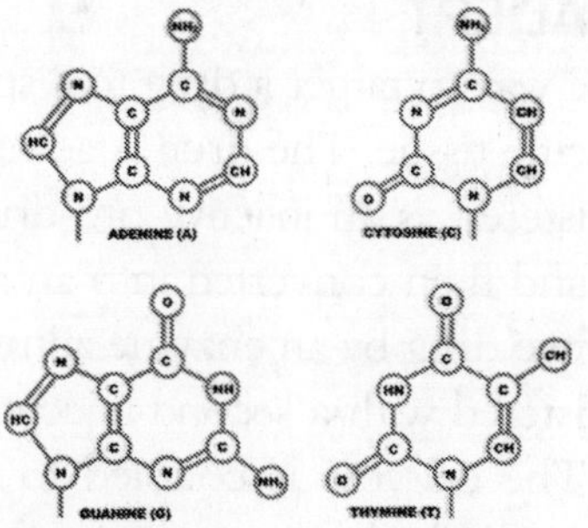

■ **adenovirus**

a group of DNA viruses which cause diseases in animals. In man, they produce acute respiratory tract infections with symptoms resembling common cold. They are used in gene cloning, as vectors for expressing large amounts of recombinant proteins in animal cells. They are also used to make live-virus vaccines against more dangerous pathogens. *See* viral vaccines.

■ **adenylate cyclase**

the enzyme that catalyses the formation of cyclic AMP.

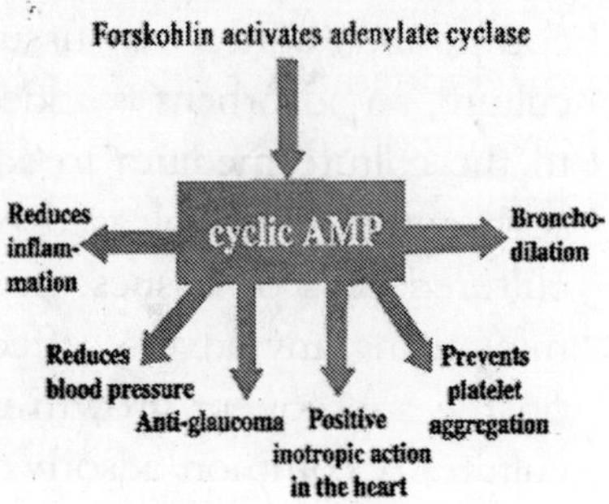

ADEPT

a way to target a drug to a specific tissue. The drug is administered as an inactive pro-drug, and then converted into an active drug by an enzyme administered with a second injection. The enzyme is coupled to an antibody that concentrates it in the target tissue. When the enzyme arrives at the target tissue, the pro-drug is activated to form the active drug, while elsewhere it remains inactive.

adhesion

the attraction of dissimilar molecules for each other. A sticking together of unlike substances, such as soil and water.

A-DNA

a right-handed DNA double helix that has 11 base pairs per turn. DNA exists in this form when partially dehydrated.

adsorbent

noun: a substance to which compounds adhere. In tissue culture, an adsorbent is added to the culture medium to adsorb compounds released by cultured cells or tissues, thus minimizing any adverse effect on the subsequent growth in culture. A common adsorbent in tissue culture is *activated charcoal*, q.v.

adsorption

the formation of a layer of gas, liquid or solid on the surface of a solid. *cf* absorption.

adult cloning

the creation of identical copies of an adult animal by nuclear transfer (q.v.) from differentiated adult tissue.

advanced

applied to an organism or a part thereof, implying considerable development from the ancestral stage or from the explant stage.

adventitious

a structure arising at sites other than the usual ones, e.g., shoots from roots or leaves, and embryos from any cell other than a zygote.

aerate

to supply with or mix with air or gas. The process is aeration.

aerobe

a micro-organism that grows in the presence of oxygen. Opposite: anaerobe.

aerobic

active in the presence of free oxygen.

■ aerobic bacteria
bacteria that can live in the presence of oxygen.

■ aerobic respiration
a type of respiration in which foodstuffs are completely oxidised to carbon dioxide and water, with the release of chemical energy, in a process requiring atmospheric oxygen.

■ affinity chromatography
a method for separating molecules by exploiting their ability to bind specifically to other molecules. There are several types of biological affinity chromatography. A biological molecule can be immobilised and a smaller molecule (ligand, q.v.,) to which it is to bind can be stuck to it, or the smaller ligand can be immobilised and the macromolecule stuck to it. A variant is to use an antibody as the immobilised molecule and use it to 'capture' its antigen: this is often called immuno-affinity chromatography.

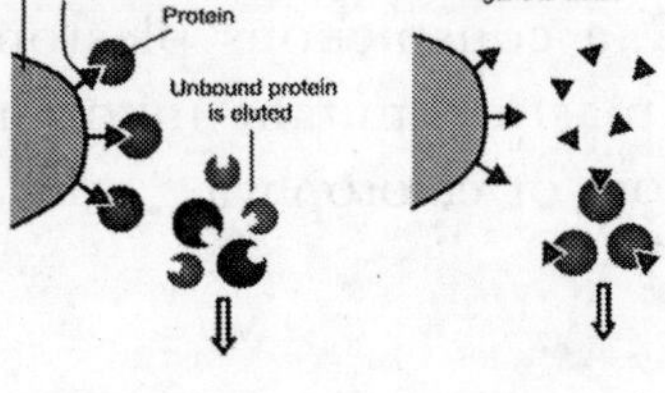

■ aflatoxin
toxic compounds, produced by moulds (fungi) of the *Aspergillus flavus* group, that bind to DNA and prevent replication and transcription. Aflatoxins can cause acute liver damage and cancer. Animals may be poisoned by eating stored food or feed contaminated with the mould.

■ agar
a polysaccharide solidifying agent used in nutrient media preparations and obtained from certain types of red algae (*Rhodophyta*). Both the type of agar and its concentration can affect the growth and appearance of cultured explants.

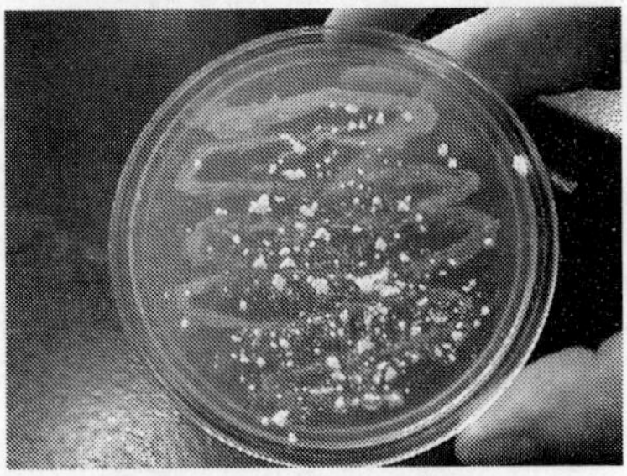

■ agarose gel electrophoresis
a process in which a matrix composed of a highly purified

form of agar is used to separate larger DNA and RNA molecules. *See* electrophoresis.

■ agonist

a drug, hormone or transmitter substance that forms a complex with a receptor site that is capable of triggering an active response from a cell.

■ agrobacterium

a genus of bacteria that includes several plant pathogenic species, causing tumour-like symptoms.

■ agrobacterium tumefaciens-mediated transformation

a naturally occurring process of DNA transfer from the bacterium *A. tumefaciens* to plants.

■ agrobiodiversity

that component of biodiversity that is relevant to food and agriculture production. The term agrobiodiversity encompasses within-species, species and ecosystem diversity. (Based on FAO, 1999).

■ AIDS

the usually fatal human disease in which the immune system is destroyed by a retrovirus (Human Immunodeficiency Virus, HIV). The virus infects and destroys helper T-cells, which are essential for combating infections.

■ airlift fermenter

a cylindrical fermentation vessel in which the cells are mixed by air introduced at the base of the vessel and that rises through the column of culture medium. The cell suspension circulates around the column as a consequence of the gradient of air bubbles in different parts of the reactor.

■ albinism

hereditary absence of pigment in an organism. Albino animals have no colour in their skin, hair and eyes. The term is also used for absence of chlorophyll in plants.

■ albino

1. an organism lacking pigmentation, due to genetic factors. The condition is *albinism*, q.v.

2. a conspicuous plastome (plastid) mutant involving loss of chlorophyll.

■ aleurone

the outermost layer of the endosperm in a seed.

■ algal biomass

single-celled plants, such as *Chlorella* spp. and *Spirulina* spp., are grown commercially in ponds to make feed materials. Chlorella is grown commercially to make into fish food: it is fed to zooplankton, and these in turn are harvested as feed for fish farms. This is a means of converting sunlight into food in a way more convenient and controllable than normal farming.

■ alginate

polysaccharide gelling agent.

■ alkylating agents

chemicals that transfer alkyl (methyl, ethyl, etc.) groups to the bases in DNA.

■ allele

one of a pair, or series, of variant forms of a gene that occur at a given locus in a chromosome. Alleles are symbolised with the same basic symbol (e.g., *B* for dominant and *b* for recessive); B_1, B_2, ..., B_n *for n* additive alleles at a locus). In a normal diploid cell there are two alleles of any one gene (one from each parent), which occupy the same relative position (locus) on homologous chromosomes. Within a population there may be more than two alleles of a gene.

■ allele frequency

the number of copies of an allele in a population, expressed as a proportion of the total number of copies of all alleles at a locus in a population.

■ allele-specific amplification (ASA)

the use of polymerase chain reaction (PCR) at a sufficiently high stringency that only a primer with exactly the same sequence as the target DNA will be amplified. A powerful means of genotyping for single-locus disorders that have been characterised at the molecular level.

■ allelic exclusion

a phenomenon whereby only one functional allele of an antibody gene can be assembled in a given B lymphocyte. The "allele" on the other homologous chromosome in a diploid mammalian cell cannot undergo a functional re-arrangement, which would result in the production of two different antibodies by a single plasma cell.

■ **allelopathy**

the phenomenon by which the secretion of chemicals, such as phenolic and terpenoid compounds, by a plant inhibits the growth or reproduction of other plant species with which it is competing.

■ **allergen**

an antigen that provokes an immune response.

■ **allogamy**

cross fertilisation in plants..

■ **allometric**.

when the growth rate of one part of an organism differs from that of another part or of the rest of the body.

■ **allopatric speciation**

speciation occurring at least in part because of geographic isolation.

■ **allopolyploid**

a polyploid organism (usually a plant) having multiple sets of chromosomes derived from different species. Hybrids are usually sterile, because they do not have sets of homologous chromosomes and therefore pairing cannot take place. However, if doubling of the chromosome number occurs in a hybrid derived from two diploid (2n) species, the resulting tetraploid (4n) is a fertile plant, since it contains two sets of homologous chromosome and pairing may occur; this tetraploid is an allotetraploid.

■ **allosteric enzyme**

an enzyme that has two structurally distinct forms, one of which is active and the other inactive. Active forms of allosteric enzymes tend to catalyse the initial step in a pathway leading to the synthesis of molecules. The end product of this synthesis can act as a feedback inhibitor, converting the enzyme to the inactive form, thus controlling the amount of product synthesised.

■ **allosteric regulation**

a catalysis-regulating process in which the binding of a small effector molecule to one site on an enzyme affects the activity at another site.

■ **allosteric transition**

a reversible interaction of a small molecule with a protein molecule, resulting in a change in the shape of the protein and a consequent alteration of the interaction of that protein with a third molecule.

■ **allotetraploid**

an organism with four genomes derived from hybridisation of different species. Usually, in forms that become established, two of the four genomes are from one species and two are from another species..

■ **allozygote**

a diploid individual that is homozygous at a locus in which the two genes are not identical by descent from a common ancestor.

■ **alphalactalbumin**

protein component of milk.

■ **alternative mRNA splicing**

the inclusion or exclusion of different exons to form different mRNA transcripts. *See* RNA.

■ **ambient temperature**

air temperature at a given time and place; not radiant temperature.

■ **amino acid**

an acid containing the group NH_2. In particular, any of 20 basic building blocks of proteins with a free amino (NH_2) and a free carboxyl (COOH) group, and having the basic formula NH_2 - C_R - COOH. According to the side group R, they are subdivided into: polar or hydrophilic (serine, threonine, tyrosine, asparagine and glutamine); non-polar or hydrophobic (glycine, alanine, valine, leucine, isoleucine, proline, phenylalanine, tryptophan and cysteine); acidic (aspartic acid and glutamic acid) and basics (lysine, arginine, hystidine). The sequence of amino acids determines the shape, properties and the biological role of a protein. Plants and many micro-organisms can synthesise amino acids from simple inorganic compounds, but animals are unable to synthesise some of them, called essential amino acids, so they must be present in the diet.

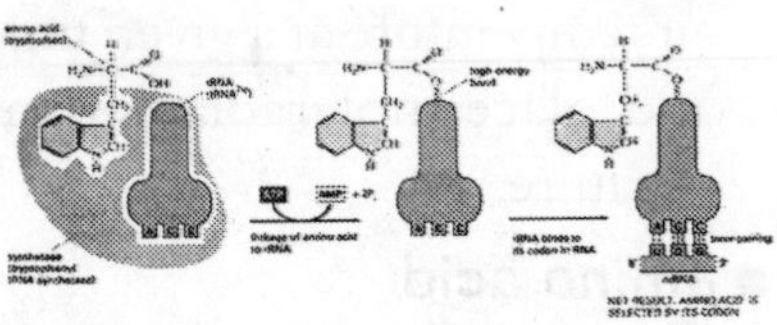

■ **aminoacyl site, A-site**

one of two sites on ribosomes to which the incoming aminoacyl tRNA binds.

■ **aminoacyl tRNA synthetase**

enzyme that attaches each amino acid to its specific tRNA molecule.

■ **amitosis**

cell division (cytokinesis), including nuclear division through constriction of the nucleus, without chromosome differentiation as in mitosis. The maintenance of genetic integrity and diploidy during amitosis is uncertain. This process occurs in the endosperm of flowering plants.

■ **amniocentesis**

a procedure for obtaining amniotic fluid from a pregnant mammal for the diagnosis of some diseases in the unborn foetus. Cells are cultured, and metaphase chromosomes are examined for irregularities (e.g., Down syndrome, spina bifida, etc., in humans).

■ **amnion**

the thin membrane that lines the fluid-filled sac in which the embryo develops in higher vertebrates, reptiles and birds.

■ **amniotic fluid**

liquid contents of the amniotic sac of higher vertebrates, containing cells of the embryo (not of the mother). Both fluid and cells are used for diagnosis of genetic abnormalities in the embryo or foetus.

■ **amorph**

a mutation that obliterates gene function.

■ **amphidiploid**

a species or type of plant derived from doubling the chromosomes in the F_1 hybrid of two species; an allopolyploid. In an amphidiploid the two species are known, whereas in other allopolyploids they may not be known.

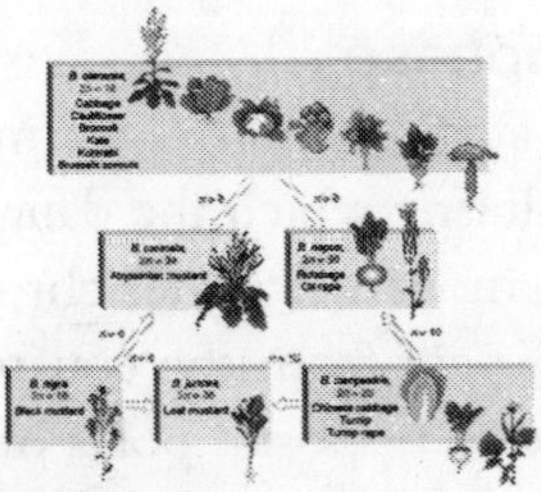

■ **amphimixis**

true sexual reproduction involving the fusion of male and female gametes and the formation of a zygote.

■ **ampicillin**

a penicillin-derived antibiotic that prevents bacterial growth by interfering with synthesis of the cell wall.

■ **amplification**

1. treatment (e.g., use of chloramphenicol) designed to increase the proportion of plasmid DNA relative to that of bacterial (host) DNA.
2. replication of a gene library in bulk.
3. duplication of gene(s) within a chromosomal segment.
4. creation of many copies of a segment of DNA by the polymerase chain reaction (PCR).

■ **amplify**

to increase the number of copies of a DNA sequence, either *in vivo* by inserting into a cloning vector that replicates within a host cell, or *in vitro* by polymerase chain reaction (PCR).

■ **amylase**

a group of enzymes that degrade starch, glycogen and other polysaccharides, producing a mixture of glucose and maltose. Plants have both a- and b-amylase; animal have only a-amylase.

■ **amylolytic**

the capability of breaking down starch into sugars.

■ **amylopectin**

a polysaccharide comprising highly branched chains of glucose molecules. The water-insoluble portion of starch.

■ **amylose**

a polysaccharide consisting of linear chains of 100 to 1000 glucose molecules. The water-soluble portion of starch.

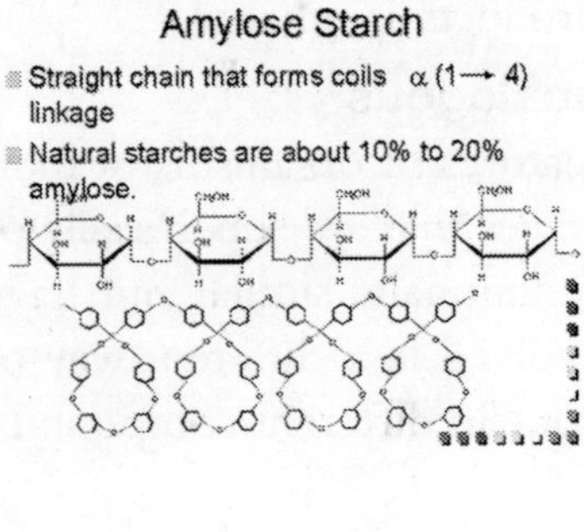

■ **anabolic pathway**

a pathway by which a metabolite is synthesised; a biosynthetic pathway.

■ **anaerobe**

an organism that can grow in the absence of oxygen. Opposite: aerobe.

■ **anaerobic**

an environment or condition in which molecular oxygen is not available for chemical, physical or biological processes.

■ **anaerobic digestion**

digestion of materials in the absence of oxygen. *See* anaerobic respiration.

■ **anaerobic respiration**

respiration in which foodstuffs are partially oxidised, with the release of chemical energy, in a process not involving atmospheric oxygen, such as alcoholic fermentation, in which one of the end products is ethanol.

■ **analogous**

features of organisms or molecules that are superficially or functionally similar but have evolved in a different way or contain different compounds.

■ **anaphase**

the stage of mitosis or meiosis during which the daughter chromosomes (sister chromatids) pass from the equatorial plate to opposite poles of the cell (toward the ends of the spindle). Anaphase follows metaphase and precedes telophase.

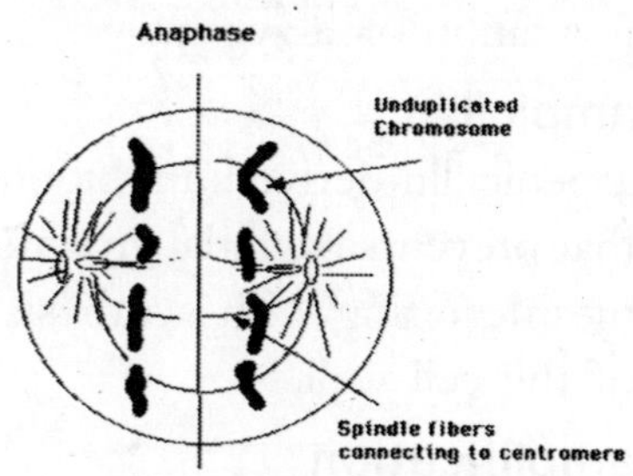

■ **anchor gene**

a gene that has been positioned on both the physical map and the linkage map of a chromosome.

■ **androgen**

any hormone that stimulates the development of male secondary sexual characteristics, and contributes to the control of sexual activity in vertebrate animals. Usually synthesised by the testes.

■ **androgenesis**

male parthenogenesis, i.e., the development of a haploid embryo from a male nucleus. The maternal nucleus is eliminated or inactivated subsequent to fertilisation of the ovum, and the haploid individual (referred to as androgenetic) contains in its cells the genome of the male gamete only. Androgenesis is detected by cytological staining..

■ **aneuploid**

an organism or cell having a chromosome number that is not an exact multiple of the monoploid *(x)* with one chromosome being present in greater (e.g., trisomic $2n + 1$) or lesser (e.g., monosomic $2n - 1$) number than the normal diploid number.

■ **animal cell immobilisation**

entrapment of animal cells in some solid material in order to produce some natural product or genetically engineered protein. Animal cells have the advantage that they already produce many proteins of pharmacological interest, and that genetically engineered proteins are produced by them with the post-translation modifications normal to animals. However, because animal cells are much more fragile than bacterial ones, they cannot tolerate a commercial fermentation process. Typical materials are hollow fibre membrane bioreactors, or porous carriers made of polysaccharide, protein, plastic or ceramic materials with microscopic holes inside which the cells grow.

■ **animal genome (gene) bank**

a planned and managed repository containing animal genetic resources. Repositories include the environment in which the genetic resource has developed, or is now normally found *(in situ)* or facilities elsewhere *(ex situ - in vivo* or *in vitro)*. For *in vitro, ex situ* genome bank facilities, germplasm is stored in the form of one or more of the following: semen, ova, embryos and tissue samples.

■ **anneal**

the pairing of complementary DNA or RNA sequences, via hydrogen bonding, to form a double-stranded polynucleotide. Most often used to describe the binding of a short primer or probe.

■ **annealing**

the process of heating (denaturing step) and slowly cooling (re-naturing step) double-stranded DNA to allow the formation of hybrid DNA or complementary strands of DNA or of DNA and RNA.

■ **annual**

1.(adj:) taking one year, or occurring at intervals of one year.

2 noun: in botany, a plant that completes its life cycle within one year. During this time the plant germinates, grows, flowers, produces seeds, and dies.

■ **anonymous DNA marker**

a DNA marker (q.v.) detectable by virtue of variation in its sequence, irrespective of whether or not it actually occurs in or near a coding sequence. Microsatellites are typical anonymous DNA markers.

■ **antagonism**

an interaction between two organisms (e.g., moulds or bacteria) in which the growth of one is inhibited by the other. *cf* synergism.

■ **antagonist**

a compound that inhibits the effect of an agonist in such a way that the combined biological effect of the two becomes smaller than the sum of their individual effects.

■ **anther**

microsporangium bearing microspores which develop into pollen (microgametophytes). The upper part of a stamen, containing pollen sacs within which are numerous pollen grains.

■ **anther culture**

the aseptic culture of anthers for the production of haploid plants from microspores.

■ **anthesis**

the flowering period or efflorescence. Anthesis is the time of full bloom, which lasts till fruit set.

■ **anthocyanin**

water-soluble blue, purple and red flavonoid pigments found in vacuoles of cells.

■ antiauxin

a chemical that interferes with the auxin response. Antiauxin may or may not involve prevention of auxin transport or movement in plants. Some antiauxins are said to promote morphogenesis *in vitro,* such as 2,3,5-tri-iodobenzoate (TIBA; f.w. 499.81), or 2,4,5-trichlorophenoxyacetate (2,4,5-T; m.w. 255.49), which stimulate the growth of some cultures.

■ antibiosis

the prevention of growth or development of an organism by a substance or another organism.

■ antibiotic

a class of natural and synthetic compounds that inhibit the growth of or kill some micro-organisms. Antibiotics such as penicillin arc often used to control (to some extent kill) contaminating organisms. However, resistance to particular antibiotics can be acquired through mutations. Some contaminating organisms are only suppressed or their metabolism slowed to an insignificant level.

■ antibiotic resistance

the ability of a micro-organism to produce a protein that disables an antibiotic or prevents transport of the antibiotic into the cell.

■ antibody

an immunological protein (called an immunoglobulin, Ig) produced by certain white blood cells (lymphocytes) of the immune system of an organism in response to a contact with a foreign substance (antigen). Such an immunological protein has the ability of specifically binding with the foreign substance and rendering it harmless. The basic immunoglobulin molecule consists of two identical heavy and two identical light chains..

■ antibody class

the class to which an antibody belongs, depending on the type of heavy chain present. In mammals, there are five classes of antibodies: IgA, IgD, IgE, IgG, and IgM.

■ antibody structure

antibodies have a well-defined structure. Each antibody has two identical "light" chains and two identical "heavy" chains. Each chain comprises a constant region, i.e., a region that is the same between antibodies of the same class and sub-class, and a variable region that differs between. The antigen-binding region or binding site - *complementarity determining region* - is in the variable region. The antibody can be cut by proteases into several fragments, known as Fab, Fab', and Fc.

■ antibody-mediated (humoral) immune response

the synthesis of antibodies by B cells in response to an encounter of the cells of the immune system with a foreign immunogen.

■ anticlinal

the plane of cell wall orientation or cell division perpendicular (at right angles) to the surface of an organ.

■ anticoding strand

the strand of the DNA double helix that is actually transcribed. Also known as the antisense or template strand.

■ anticodon

a triplet of nucleotides in a tRNA molecule that pairs with a complementary triplet of nucleotides, or codon, in an mRNA molecule during translation.

■ antigen; immunogen

a compound that elicits an immune response by stimulating the production of antibodies. The antigen, usually a protein, when introduced into a vertebrate organism is bound by the antibody or a T cell receptor.

■ antigenic determinant

a surface feature of a microorganism or macromolecule, such as a glycoprotein, that elicits an immune response.

■ antigenic switching

the altering of a microorganism's surface antigens through genetic re-arrangement, to elude detection by the host's immune system.

■ antihaemophilic globulin
blood globulin that reduces the clotting time of haemophilic blood.

■ anti-idiotype antibodies
antibodies which recognise the binding sites of other antibodies. Their binding sites are complementary to the binding sites of another immunoglobulin. When an animal becomes immune to something, it not only acquires antibodies against that something, it also acquires antibodies against those antibodies. This forms a network of antibodies which can all bind to each other to various degrees, helping to regulate the immune response. Some allergic responses are in part due to the breakdown of this sort of regulation.

■ antimicrobial agent
any chemical or biological agent that harms the growth of micro-organisms.

■ antinutrients
compounds that inhibit normal uptake of nutrients.

■ anti-oncogene
a gene whose product prevents the normal growth of tissue. *cf* recessive oncogene.

■ antioxidant
compound that slows the rate of oxidation reactions.

■ antioxidant solution
pre-treatment solution (e.g., Vitamin C; citric acid) that retards senescence and browning of tissue. It is employed to incubate explants prior to surface sterilisation.

■ antiparallel orientation
the normal arrangement of the two strands of a DNA molecule, and of other nucleic-acid duplexes (DNA-RNA, RNA-RNA), in which the two strands are oriented in opposite directions so that the 5´-phosphate end of one strand is aligned with the 3´-hydroxyl end of the complementary strand.

■ antisense DNA
1. the strand of chromosomal DNA that is transcribed.
2. a DNA sequence that is complementary to all or part of an mRNA molecule.

■ antisense gene

a gene that produces a transcript (mRNA) that is complementary to the pre-mRNA or mRNA of a normal gene (usually constructed by inverting the coding region relative to the promoter).

■ antisense RNA

an RNA sequence that is complementary to all or part of a functional mRNA molecule, to which it binds, blocking its translation. *See* RNA.

■ antisense therapy

the *in vivo* treatment of a genetic disease by blocking translation of a protein with a DNA or an RNA sequence that is complementary to a specific mRNA.

■ antiseptic

any substance that kills or inhibits the growth of disease-causing micro-organism (a micro-organism capable of causing sepsis), but is essentially non-toxic to cells of the body.

■ antiserum

the fluid portion of the blood of an animal (after coagulation of the blood), containing antibodies.

■ anti-terminator

a type of protein which enables RNA polymerase to ignore certain transcriptional stop or termination signals and read through them to produce longer mRNA transcripts.

■ antitranspirant

a compound designed to reduce transpiration when sprayed or painted on leaves of newly transplanted trees, shrubs or vines, or used as a dip for cuttings in lieu of misting; may interfere with photosynthesis and respiration if the coating is too thick or unbroken.

■ apex

the tip, point or angular summit. The tip of a leaf; the portion of a root or shoot containing apical and primary meristems. Usually used to designate the apical tip of the meristem.

■ apical cell

a meristematic initial in the apical meristem of shoots or roots of plants. As this cell divides, new tissues are formed.

■ **apical dominance**

the phenomenon of inhibition of growth of lateral (axillary) buds in a plant by the presence of the terminal (apical) bud on the branch, due to auxins produced by the apical bud.

■ **apical meristem**

a region of the tip of each shoot and root of a plant in which cell division is continually occurring to produce new stem and root tissue, respectively. Two regions are visible in the apical meristem: (i) An outer 1-4-cell layered region (called the tunica), where cell divisions are anticlinal, i.e., perpendicular to the surface; and below the tunica, (ii) the corpus, where the cells divide in all directions, giving them an increase in volume.

■ **apoenzyme**

inactive enzyme that has to be associated with a specific organic molecule called a co-enzyme in order to function. The apoenzyme/co-enzyme complex is called a holoenzyme.

■ **apomixis**

the asexual production of diploid offspring without the fusion of gametes. The embryo develops by mitotic division of the maternal or paternal gamete, or, in the case of plants, by mitotic division of a diploid cell of the ovule. *cf* androgenesis; gynogenesis; panmixis; parthenogenesis.

■ **apoptosis**

the process of cell death, which occurs naturally as a part of normal development, maintenance and renewal of tissue in an organism. Apoptosis differs from necrosis, in which cell death is caused by a toxic substance.

■ **aquaculture**

growing of water plants and animals, rather than harvesting them from wherever they happen to grow in rivers or seas. Usually aquaculture uses fresh water; when it uses sea water it can be called mariculture. It is considered to be a part of biotechnology (although peripheral) because it is a new commercial development, and because it often involves growing organisms in large volumes of water, which has similarities to growing large volumes of yeast or bacteria. Biotechnology also provides clean, well-aerated water for the animals to grow in; food, such as krill or powdered synthetic food; and food additives, such as astaxanthins to ensure that fish and prawns have the right colour. Aquaculture has also been used to mass-produce macro- and micro-algae for chemicals, vitamins and pigments. For both animals and plants, biotechnologists have been using genetic methods to produce triploid and tetraploid organisms, and hybrid algae through plant cell fusions. Triploid trout, for example, are sterile, and can be used for biocontrol of weeds without the threat of their being able to breed themselves.

■ **artificial inembryonation**

non-surgical transfer of embryo(s) to a recipient female. As *in vitro* embryo technology develops, artificial inembryonation will gradually replace artificial insemination.

■ **artificial insemination**

the deposition of semen, using a syringe, at the mouth of the uterus to make conception possible. It is used in the breeding of domestic animals.

■ **artificial seed**

encapsulated or coated somatic embryos (embryoids) that are planted and treated like seed.

■ **artificial selection**

the practice of choosing individuals from a population for reproduction, usually because these individuals possess one or more desirable traits.

■ **ascorbic acid**

vitamin C, a water-soluble vitamin present naturally in some plants, and also synthetically produced. Aside from its role as a vitamin, it is used as an antioxidant in plant tissue culture; and included in disinfection solutions.

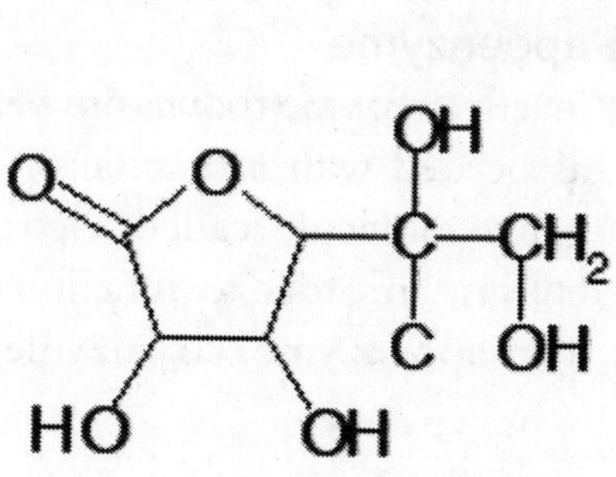

■ **ascospore**

one of the spores contained in the ascus (q.v.) of certain fungi.

■ **ascus**

reproductive sac in the sexual stage of a type of fungi (*Ascomycetes*) in which ascospores are produced.

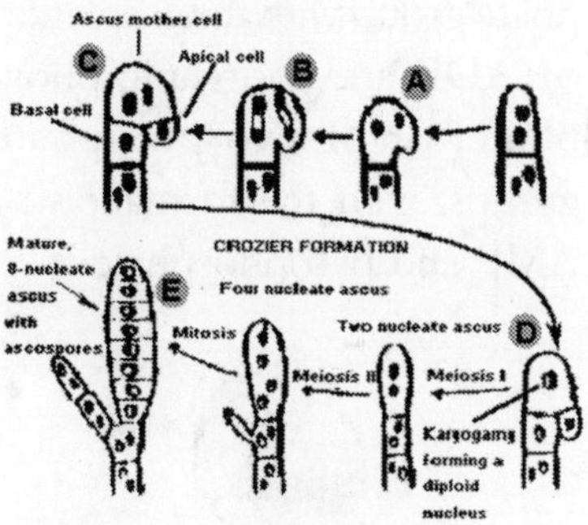

■ **aseptic**

asepsis or sterile. The state of being free of contaminating organisms (bacteria, fungi, algae and all micro-organisms except viruses) but not necessarily free of internal symbionts.

■ **asexual**

any type of reproduction not involving meiosis or the union of gametes.

■ **asexual embryogenesis**

the sequence of events whereby embryos develop from somatic cells. a.k.a. somatic cell embryogenesis.

■ **asexual propagation**

vegetative, somatic, non-sexual reproduction of a plant without fertilisation. *cf* apomixis.

■ **asexual reproduction**

reproduction that does not involve the formation and union of gametes from the different sexes or mating types. It occurs mainly in lower animals, micro-organisms and plants. In plants, asexual reproduction is by vegetative propagation (e.g., bulbs, tubers, corms) and by formation of spores.

■ **asparagine**

one of the 20 essential amino acids. It is occasionally included in plant tissue culture media as a source of reduced nitrogen.

■ **aspartic acid**

an amino acid necessary for nucleotide synthesis and occasionally included in plant tissue culture media.

■ **assay**

1.to test or evaluate.

2. the procedure for measuring the quantity of a given substance in a sample (chemically or by other means). The substance to be analysed.

■ assortative mating

mating in which the partners are chosen on the basis of phenotypic similarity.

■ asynapsis

the failure or partial failure in the pairing of homologous chromosomes during the meiotic prophase.

■ ATP

a nucleotide of fundamental importance as a carrier of chemical energy in all living organisms. It consists of adenosine with three phosphate groups, linked together linearly. The phosphates are attached to adenosine through its ribose (sugar) portion. Upon hydrolysis, bonds yield either one molecule of ADP (adenosine diphosphate) and an inorganic phosphate, or one molecule of AMP (adenosine monophosphate) and pyrophosphate; in both cases releasing energy that is used to power biological processes. ATP is regenerated by rephosphorilation of AMP and ADP, using chemical energy derived from the oxidation of food.

■ ATP-ase

an enzyme that brings about the hydrolysis of ATP, by the cleavage of either one phosphate group with the formation of ADP and inorganic phosphate, or of two phosphate groups, with the formation of AMP and pyrophosphate.

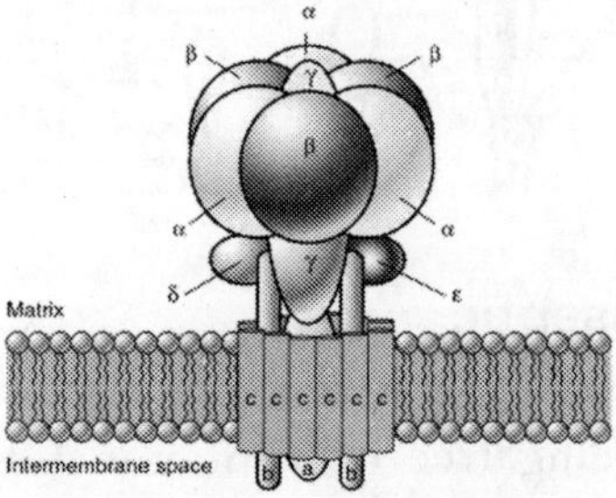

■ attenuated vaccine

a virulent organism that has been modified to produce a less virulent form, but nevertheless retains the ability to elicit antibodies against the virulent form.

■ attenuation

a mechanism for controlling gene expression in prokaryotes that involves premature termination of transcription.

■ attenuator

a nucleotide sequence in the 5′ region of a prokaryotic gene (or in its RNA) that causes premature termination of transcription, possibly by forming a secondary structure.

■ authentic protein

a recombinant protein that has all the properties - including any post-translational modifications - of its naturally occurring counterpart.

■ autocatalysis

catalysis in which one of the products of the reaction is a catalyst for the reaction. Usually the catalysis starts slowly and increases as the quantity of the catalyst increases, falling off as the product is used up.

■ autoclave

1.an enclosed chamber in which substances are heated under pressure to sterilise utensils, liquids, glassware, etc., using steam. The routine method uses steam pressure of 103.4×10^3 Pa at 121°C for 15 minutes, or longer to allow large volumes to reach the critical temperature.

2. a pressure cooker used to sterilise growth medium and instruments for tissue culture work.

■ auto-immune disease

disorder in which the immune systems of affected individuals produce antibodies against molecules that are normally produced by those individuals (called self antigens).

■ auto-immunity

a disorder in the body's defence mechanism in which an immune response is elicited against its own (self) tissues.

■ autologous cells

cells that are taken from an individual, cultured (or stored), and, possibly, genetically manipulated before being infused back into the original donor.

■ autolysis

the process of self destruction of a cell, cell organelle, or tissue. It occurs by the action of lysosomic enzymes.

■ autonomous

a term applied to any biological unit that can function on

its own, i.e., without the help of another unit, such as a transposable element that encodes an enzyme for its own transposition.

■ **autopolyploid**

a polyploid that has multiple and identical or nearly identical sets of chromosomes (genomes) all derived from the same species. A polyploid species with genomes derived from the same original species.

■ **autoradiograph**

a picture prepared by labelling a substance such as DNA with a radioactive material such as tritiated thymidine and allowing the image produced by decay radiation to develop on a film over a period of time.

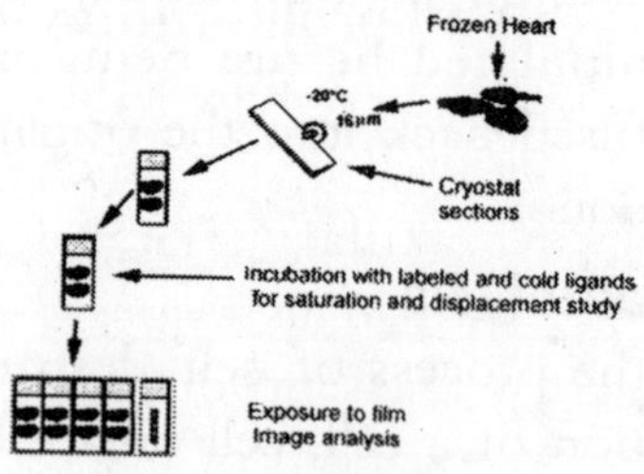

■ **autoradiography**

a technique that captures the image formed in a photographic emulsion as a result of the emission of either light or radioactivity from a labelled component that is placed next to unexposed film. The technique is used for detecting the location of an isotope in a tissue, cell or molecule. The sample is placed in contact with a photographic emulsion, usually an X-ray film. The emission of b-particles from the sample activates the silver halide grains in the emulsion and allows them to reduce to metallic silver when the film is developed. In genetic engineering, autoradiography is most commonly used to detect the hybridisation of a radioactive DNA (probe) molecule to denatured DNA in either the Southern transfer or colony hybridisation procedures.

■ **autosome**

a chromosome that is not involved in sex determination.

■ **auxin**

a group of plant growth regulators (natural or synthetic) which stimulate cell division, enlargement, apical dominance, root initiation, and flowering. One naturally produced auxin is indole-acetic acid (IAA).

■ **auxin-cytokinin ratio**
the relative proportion of auxin to cytokinin present in plant-tissue-culture media. Varying the relative amounts of these two hormone groups in tissue culture formulae affects the proportional growth of shoots and roots *in vitro*. As the ratio is increased (increased auxin or decreased cytokinin), roots are more likely to be produced, and as it is decreased root growth declines and shoot initiation and growth are promoted. This relationship was first recognised by C.O. Miller and F. Skoog in the 1950s.

■ **auxotroph**
a mutant cell or micro-organism lacking the capacity to form an enzyme or metabolite present in the parental strain, and that consequently will not grow on a minimal medium, but requires the addition of some compound - such as an amino acid or a vitamin - for growth.

■ **axenic culture**
free of external contaminants and internal symbionts; generally not possible with surface sterilisation alone, and incorrectly used to indicate *aseptic culture*, q.v.

■ **axillary bud**
a bud found at the axil of a leaf (synonymous with lateral bud).

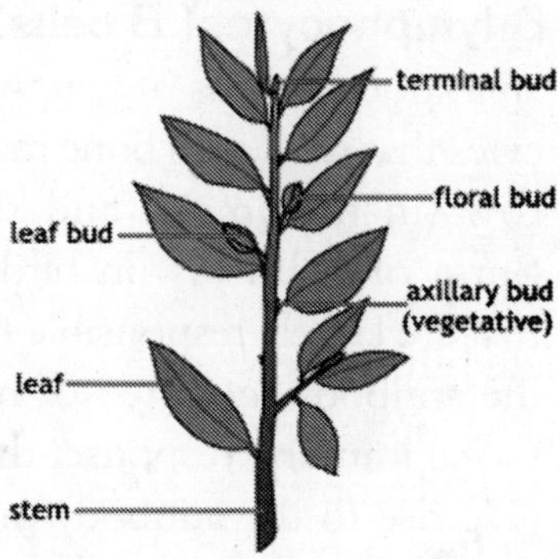

■ **axillary bud proliferation**
propagation in culture using protocols and media which promote axillary (lateral shoot) growth. This is a technique for mass production (micropropagation) of plantlets in culture, achieved primarily through hormonal inhibition of apical dominance and stimulation of lateral branching.

■ **B cells**

an important class of white blood cells that mature in bone marrow and produce antibodies. They are largely responsible for the antibody-mediated or humoral immune response; they give rise to the antibody-producing plasma cells and some other cells of the immune system. *See* B lymphocytes.

■ **B lymphocytes(B cells)**

an important class of lymphocytes that mature in bone marrow (in mammals) and the Bursa of Fabricius (in birds), that are largely responsible for the antibody-mediated or humoral immune response; they give rise to the antibody-producing plasma cells and some other cells of the immune system.

■ **BAC**

a cloning vector constructed from bacterial fertility (F) factors; like YAC vectors, they accept large inserts of size 200 to 500 kb.

■ **bacillus**

a rod-shaped bacterium.

■ **bacillus thuringiensis**

a bacterium that kills insects; a major component of the microbial pesticide industry.

■ **back mutation**

a second mutation at the same site in a gene as the original mutation. The second mutation restores the wild-type nucleotide sequence.

■ **backcross**

crossing an organism with one of its parents or with the genetically equivalent organism. The offspring of such a cross are referred to as the backcross generation or backcross progeny.

■ **bacterial toxin**

a toxin produced by a bacterium, such as Bt toxin by *Bacillus thuringiensis*.

■ **bacteriocide**

a chemical or drug that kills bacterial cells.

■ **bacteriocin**

a protein produced by bacteria of one strain and active against those of a closely related strain.

■ **bacteriostat**

a substance that inhibits or slows down growth and reproduction of bacteria.

■ balanced lethal system

a system for maintaining a recessive lethal allele at each of two loci on the same pair of chromosomes. In a closed population with no crossing-over between the loci, only the double heterozygotes for the lethal mutations survive.

■ barr body

a condensed mass of chromatin found in the nuclei of female mammals. It is a late-replicating, inactive X-chromosome. Named after its discoverer, Murray Barr (1908-).

■ basal

1. Located at the base of a plant or a plant organ.
2. A fundamental formulation of a tissue culture medium.

■ basal body

small granule to which a cilium or flagellum is attached. *cf* kinetosome.

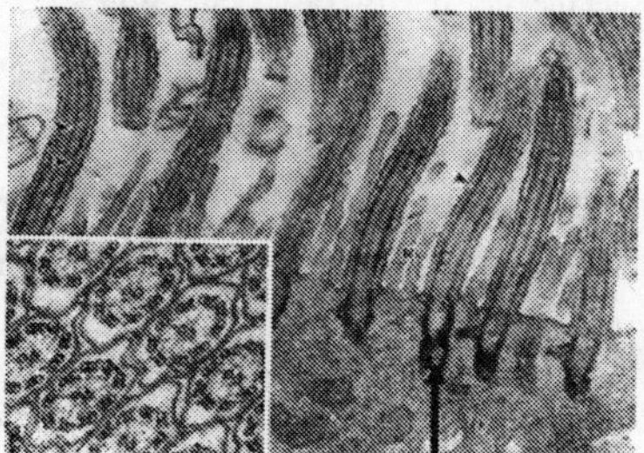

■ base

a cyclic, nitrogen-containing compound that is one of the essential components of nucleic acids. Exists in five main forms (adenine, A; guanine, G; thymine, T; cytosine, C; uracil, U). A and G have a similar structure and are called purines; T, C and U have a similar structure and are called pyrimidines. A base joined to a ribose sugar joined to a phosphate group is a nucleotide - the building block of nucleic acids.

■ base analogues

unnatural purine or pyrimidine bases that differ slightly in structure from the normal bases, but can be incorporated into nucleic acids. They are often mutagenic.

■ base collection

in PGR: Defined in the International Undertaking on Plant Genetic Resources (FAO, 1983) as a collection of seed stock or vegetative propagating material (ranging from tissue cultures to whole plants) held for long-term security in order to preserve the genetic variation for scientific purposes and as a basis for plant breeding as multiplication and evaluation. *cf* active collection.

■ **base pair**

the two strands that constitute DNA are held together by specific hydrogen bonding between purines and pyrimidines (A pairs with T; and G pairs with C). The size of a nucleic acid molecule is often described in terms of the number of base pairs (symbol: bp) or thousand base pairs (kilobase pairs; symbol: kb; a more convenient unit) it contains.

■ **base substitution**

replacement of one base by another in a DNA molecule.

■ **basophil**

a type of white blood cell (leucocyte), produced by stem cells in the red bone marrow.

Basophils

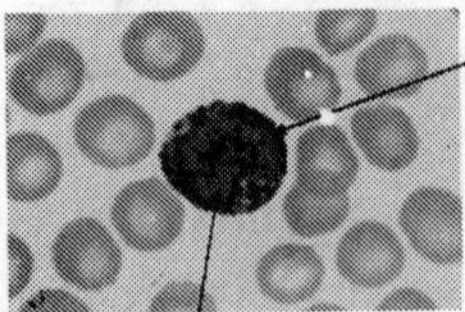

■ **batch culture**

a suspension culture in which cells grow in a finite volume of liquid nutrient medium and follow a sigmoid pattern of growth. *cf* continuous culture; batch fermentation.

■ **batch fermentation**

a process in which cells or micro-organisms are grown for a limited time. At the beginning of the fermentation, an inoculum is introduced into fresh medium, with no addition or removal of medium for the duration of the process.

■ **bench-scale process**

a small- or laboratory-scale process; commonly used in connection with fermentation.

■ **biennial**

in botany, a plant which completes its life cycle within two years and then dies. For most biennial plants, the two growing seasons have to be separated by a period of cold temperature sufficient to induce flowering and fruit formation.

■ **binary vector system**

a two-plasmid system in *Agrobacterium tumefaciens* for

transferring into plant cells a segment of T-DNA that carries cloned genes. One plasmid contains the virulence gene (responsible for transfer of the T-DNA), and another plasmid contains the T-DNA borders, the selectable marker and the DNA to be transferred.

binding

the ability of molecules to stick to each other because of the exact shape and chemical nature of parts of their surfaces. Many biological molecules bind extremely tightly and specifically to other molecules: enzymes to their substrates; antibodies to their antigens; DNA strands to their complementary strands; and so on. Binding can be characterised by a *binding constant* or *association constant* (K_a), or its inverse, the *dissociation constant* (K_d).

binomial nomenclature

in biology, each species is generally identified by two terms: the first is the genus to which it belongs, and the second is the specific epithet that distinguishes it from others in that genus (e.g., *Quercus suber*, cork oak). The genus name always has an initial capital; the specific epithet is never capitalized, even though it may be derived from a proper name (e.g., keranda nut, *Elaeocarpus bancroftii*). Both terms in the binomial are italicized. Based on the system of classification developed by Carolus Linnaeus.

bio

a prefix derived from *bios* and used in scientific words to associate the concept of "living organisms." Usually written with a hyphen before vowels, for emphasis or in neologisms; otherwise usually without a hyphen.

bio-accumulation

in an organism, concentration of materials which are not components critical for that organism's survival. Usually it refers to the accumulation of metals or other compounds (e.g., DDT). Many organisms - plants, fungi, protists, bacteria, etc. - accumulate metals when grown in a solution of them, either as part of their defence mechanism against the poisonous effect of those compounds, or as a side-effect of the chemistry of their cell walls. Bio-accumulation is important as part of the microbial mining cycle (q.v.), removing toxic metals from wastewater, as a purification (bioremediation) process, etc.

■ bio-assay

a procedure for the assessment of a substance by measuring its effect in living cells or on organisms. Animals have been used extensively in drug research in bio-assays for the pharmacological activity of drugs. However, bio-assays are now usually developed using bacteria or animal or plant cells, as these are usually much easier to handle than whole animals or plants, are cheaper to make and keep, and avoid the ethical problems associated with testing of animals. Sometimes used to detect minute amounts of substances that influence or are essential to growth.

■ bio-augmentation

increasing the activity of bacteria that decompose pollutants; a technique used in bioremediation.

■ biocatalysis

use of enzymes to catalyse chemical reactions. *See* biotransformation.

■ biocontrol

the control of living organisms (especially pests) by biological means. Any process using deliberately introduced living organisms to restrain the growth and development of other, very often pathogenic, organisms, such as the use of spider mites to control cassava mealy bug, or the introduction of myxomatosis into Australia to control rabbits. The term also applies to use of disease-resistant crop cultivars. Biotechnology approaches biocontrol in various ways, such as using fungi, viruses or bacteria which are known to attack an insect or weed pest.

■ bioconversion

conversion of one chemical into another by living organisms, as opposed to their conversion by enzymes (which is biotransformation) or by chemical processes. The usefulness of bioconversion is much the same as that of biotransformation - in particular its extreme specificity and ability to work in moderate conditions. However, bioconversion has several other properties, including the possibility of having several chemical steps. A major commercial application is in the manufacture of steroids. The "basic" steroid molecule, often isolated from plants, is itself a very complicated molecule, and not one that is easy to modify by normal chemical means to produce the very specific molecules needed for drug use. However, a particular type of bioconversion that attacks only specific bits of the molecule can be used. Bioconversion is particularly useful for introducing chemical changes at specific points in large, complex molecules.

■ **biodegradation**

the breakdown by living organisms of a compound to its chemical constituents. Materials that can be easily biodegraded are colloquially termed *biodegradable*.

■ **biodiversity**

1.the variety of species (species diversity) or other taxa of animals, micro-organisms and plants in a natural community or habitat, or of communities in a particular environment (ecological diversity), or of genetic variation in a species (genetic diversity, q.v.). The maintenance of a high level of biodiversity is important for the stability of ecosystems.

2. the variety of life in all its forms, levels and combinations, encompassing genetic diversity, species diversity and ecosystem diversity. *cf* agrobiodiversity.

■ **bio-energetics**

the study of the flow and the transformations of energy that occur in living organisms.

■ **bio-engineering**

the use of artificial tissues, organs and organ components to replace parts of the body that are damaged, lost or malfunctioning.

bio-enrichment

adding nutrients or oxygen to increase microbial breakdown of pollutants.

bio-ethics

the branch of ethics that deals with the life sciences and their potential impact on society. At one extreme, it can be enormously useful in focusing attention on problems that need to be confronted; at the other extreme, it can become a name-calling argument between the "pro-biotechnology" and "anti-biotechnology" schools of thought, which, as it reduces discussion to epithets and clichés, can make better sound bites.

biofuel

a gaseous, liquid or solid fuel that contains energy derived from a biological source. For example, rapeseed oil or fish liver oil can be used in place of diesel fuel in modified engines. A commercial application is the use of modified rapeseed oil, which - as rapeseed methyl ester (RME) - can be used in modified diesel engines, and is sometimes named bio-diesel. *cf* biogas.

biogas

a mixture of methane and carbon dioxide resulting from the anaerobic decomposition of waste such as domestic, industrial and agricultural sewage. a.k.a. *gobar*.

biogenesis

the principle that a living organism can only arise from other living organisms similar to itself and can never originate from non-living material.

bio-informatics

the use and organisation of information of biological interest. In particular, it is concerned with organising biomolecular databases, in getting useful information out of such databases, in utilising powerful computers for analysing such information, and in integrating information from disparate biological sources.

biolistics

a technique to insert DNA into cells. The DNA is mixed with small metal particles -

usually tungsten or gold - a fraction of a micrometre across. These are then fired into a cell at very high speed. They puncture the cell and carry the DNA into the cell. Biolistics has an advantage over transfection, transduction, etc., because it can apply to any cell, or indeed to parts of a cell. Thus use of biolistics has inserted DNA into animal, plant and fungal cells, and into mitochondria inside cells. a.k.a. microprojectile bombardment.

■ biomass

1. the cell mass produced by a population of living organisms.
2. the organic mass that can be used either as a source of energy or for its chemical components.
3. all the organic matter that derives from the photosynthetic conversion of solar energy.

■ biomass concentration

the amount of biological material in a specific volume.

■ biome

a major ecological community or complex of communities, extending over a large geographical area and characterised by a dominant type of vegetation.

■ biometry

the application of statistical methods to the analysis of biological problems.

■ biopesticide

a compound that kills organisms by virtue of specific biological effects rather than as a broader chemical poison. Specific types include bio-insecticides and bio-fungicides. Bio-pesticides differ from biocontrol agents in that bio-pesticides are passive agents, whereas biocontrol agents are active, seeking out the pest to be destroyed. There are some extremely attractive anti-pest materials, such as *Bacillus thuringiensis* (Bt) toxin, which specifically interferes with the absorption of food from the guts of some insects but is harmless to mammals. The rationale behind developing bio-pesticides is that they are more likely to be biodegradable and are targeted at specific elements of the pest's metabolism.

■ **biopolymer**

any large polymeric molecule (protein, nucleic acid, polysaccharide, lipid) produced by a living organism.

■ **bioprocess**

any process that uses complete living cells or their components (e.g., enzymes, chloroplasts) to effect desired physical or chemical changes.

■ **bioreactor**

a tank in which cells, cell extracts or enzymes carry out a biological reaction. Often refers to a growth chamber (fermenter, fermentation vessel) for cells or micro-organisms.

■ **bioremediation**

a process that uses living organisms to remove contaminants, pollutants or unwanted substances from soil or water. *cf* bio-augmentation; bio-enrichment.

■ **biosensor**

a device that uses an immobilised agent (such as an enzyme, antibiotic, organelle or whole cell) to detect or measure a chemical compound. A reaction between the immobilised agent and the molecule being analysed is transduced into an electric signal.

■ **biosynthesis**

synthesis of compounds by living cells, which is the essential feature of anabolism.

■ **biotechnology**

1. the use of biological processes or organisms for the production of materials and services of benefit to humankind. Biotechnology includes the use of techniques for the improvement of the characteristics of economically important plants and animals and for the development of micro-organisms to act on the environment.

2. the scientific manipulation of living organisms, especially at the molecular genetic level, to produce new products, such as hormones, vaccines or monoclonal antibodies.

■ **biotic factor**

other living organisms that are a factor of an organism's

environment, and form the biotic environment, affecting the organism in many ways.

■ biotic stress

stress resulting from living organisms which can harm plants, such as viruses, fungi, bacteria, parasitic weeds and harmful insects. *cf* abiotic stress.

■ biotin

a vitamin of the B complex. It is a co-enzyme for various enzymes that catalyse the incorporation of carbon dioxide into various compounds. It is essential for the metabolism of fats. Biotin is attached to pyruvate carboxylase by a long, flexible chain like that of lipoamide in the pyruvate dehydrogenase complex. Adequate amounts are normally produced by the intestinal bacteria in animals.

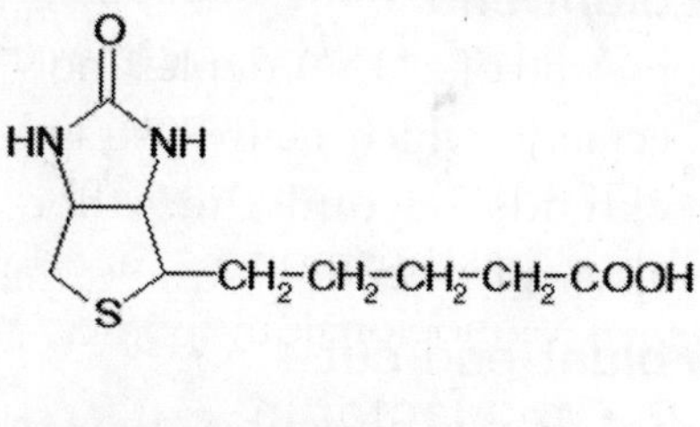

■ biotin labelling

1. the attachment of biotin to another molecule.

2. the incorporation of a biotin-containing nucleotide into a DNA molecule.

■ biotinylated-DNA

a DNA molecule labelled with biotin by incorporation of biotinylated-dUTP into a DNA molecule. It is used as a non-radioactive probe in hybridisation experiments, such as Southern transfer. The detection of the labelled DNA is achieved by complexing it with streptavidin (an antibiotic with a high affinity for biotin) to which is attached a colour-generating agent such as horseradish peroxidase that gives a fluorescent green colour upon reaction with various organic reagents.

■ biotope

a small habitat in a large community.

■ biotoxin

a naturally produced toxic compound which shows pronounced biological activity and presumably has some adaptive significance to the organism which produces it.

■ biotransformation

the conversion of one chemical or material into another using a biological catalyst: a near synonym is biocatalysis, and hence the catalyst used is called a biocatalyst. Usually the catalyst is an enzyme, or a whole, dead micro-organism that contains an enzyme or several enzymes.

■ blastocyst

a mammalian embryo (fertilised ovum) in the early stages of development, approximately up to the time of implantation. It consists of a hollow ball of cells.

■ blastomere

any one of the cells formed from the first few cleavages in animal embryology. The embryo usually divides into two, then four, then eight blastomeres, and so on.

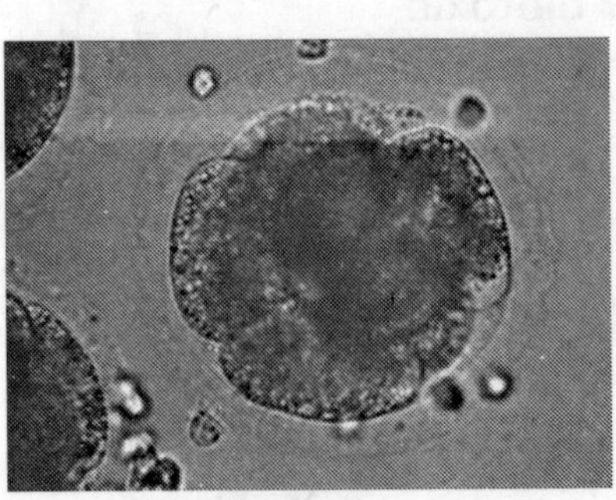

■ blastula

in animals, an early embryo form that follows the morula stage; typically, a single-layered sheet (blastoderm) or ball of cells (blastocyst).

■ bleach

a fluid, powder or other whitening (bleaching) or cleaning agent, usually with free chlorine ions. Commercial bleach contains calcium hypochlorite or sodium hypochlorite, and is a common disinfectant used for cleaning working surfaces, tools and plant materials in plant tissue culture and grafting.

■ bleeding

used to describe the occasional purplish-black coloration of media due to phenolic products given off by (usually fresh) transfers.

■ blunt end

the end of a DNA duplex molecule in which neither strand extends beyond the other. a.k.a. flush end.

■ blunt-end cut

to cleave phospho-diester bonds in the backbone of duplex DNA between the corresponding nucleotide pairs on opposite strands. This cleavage process results in both strands finishing at the same residue, i.e., there are no nucleotide extensions on either strand. a.k.a. flush-end cut.

■ blunt-end ligation

joining (ligation) of the nucleotides that are at the ends of two blunt-ended DNA duplex molecules.

■ boring platform

sterile bottom half of a Petri dish used for preparing explants with a cork borer.

■ bound water

water held by the cell and not released if freezing occurs in the intercellular space. *cf* free water.

■ bovine somatotrophin (BST)

a.k.a. bovine growth hormone, this protein is found naturally in cattle, and is the bovine counterpart of human growth hormone, one of the earliest biopharmaceutical products. It has been cloned, using recombinant DNA technology, expressed in large amounts and marketed as an agricultural product to improve the growth rate and protein:fat ratios in farm cattle, and to improve milk yield. Its use is banned in some countries.

■ bovine spongiform encephalopathy

a.k.a. mad cow disease.

■ bract

a modified leaf that subtends flowers or inflorescences and may appear to be a petal.

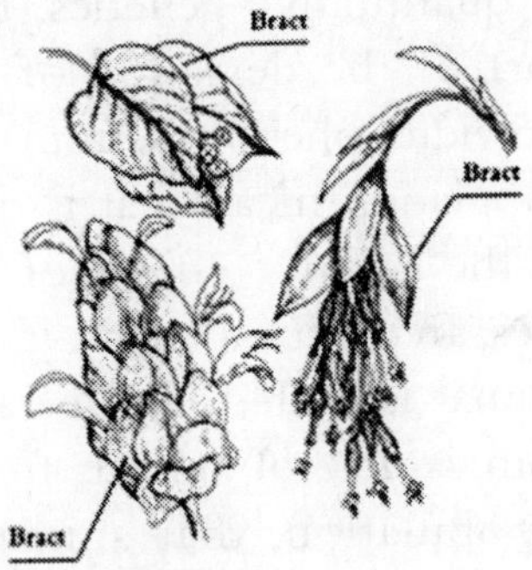

■ breed

noun: in AnGR, either (i) a sub-specific group of domestic livestock with definable and identifiable external characteristics that enable it to be separated by visual appraisal from other similarly defined groups within the same species or (ii) a group of domestic livestock for which geographical and/or cultural separation from phenotypically similar groups has led to acceptance of its separate identity. *cf* breed at risk; breed not at risk; critical breed; critical-maintained breed; endangered-maintained breed.

■ **breeding**

the process of sexual reproduction and production of offspring.

■ **breeding value**

in quantitative genetics, the part of the deviation of an individual phenotype from the population mean that is due to the additive effects of alleles. In practical terms: if an animal is mated with a random sample of animals from a population, that animal's breeding value for a certain trait is twice the average deviation of its offspring from the population mean for that trait.

■ **brewer's yeast**

strains of yeast, often *Saccharomyces cerevisiae,* that are used in the production of beer.

■ **brewing**

the process by which beer is made. In the first stage the barley grain is soaked in water and allowed to germinate (malting), during which the natural enzymes of the grain convert the seed starch to maltose, and then to glucose. Grain is then dried, crushed, and added to water at a specific temperature (steeping) and any remaining starch is converted to sugar. The resulting liquid (wort) is the raw material to which yeast is added to convert sugar to alcohol. Hops (female flowers of *Humulus lupulus*) are added during this process to give a characteristic flavour.

■ **bridge**

a filter paper or other substrate used as a wick and support structure for a plant tissue in culture when a liquid medium is used.

■ **broad-host-range plasmid**

a plasmid that can replicate in a number of different bacterial species.

■ broad-sense heritability
in quantitative genetics, the proportion of the total phenotypic variation due to genetic variation.

■ browning
discoloration due to phenolic oxidation of freshly cut surfaces of explant tissue. In later stages of culture, such discoloration may indicate a nutritional or pathogenic problem, generally leading to necrosis.

■ bubble column fermenter
a fermentation vessel, or bioreactor, in which the cells or micro-organisms are kept suspended in a tall cylinder by rising air, which is introduced at the base of the vessel.

■ bud
a region of meristematic tissue with the potential for developing into leaves, shoots, flowers or combinations; generally protected by modified scale leaves. A terminal (or apical) bud exists at the tip of a stem or branch, while axillary (or lateral) buds develop in the axils of leaves.

■ bud scar
a scar left on a shoot when the bud or bud scales drop.

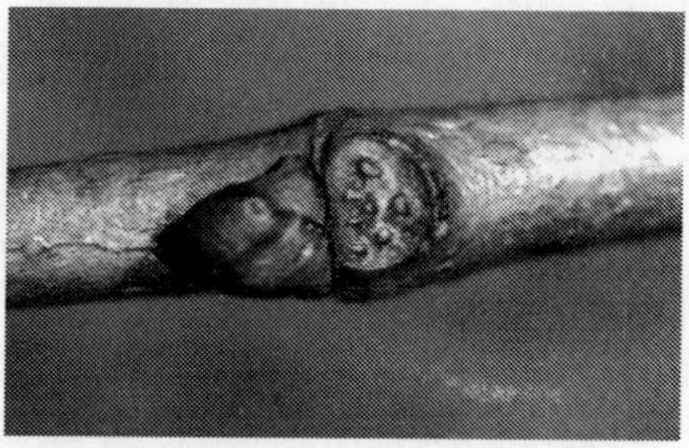

■ bud sport
a somatic mutation arising in a bud and producing a genetically different shoot. Bud sports includes changes due to gene mutation, somatic reduction, chromosome deletion or polyploidy.

■ budding
1. a method of asexual reproduction in which a new individual is derived from an outgrowth (bud) that becomes detached from the body of the parent.
2. among fungi, budding is characteristic of the yeast *Saccharomyces cerevisiae.*
3. a form of grafting in which a single vegetative bud is taken from one plant and inserted into stem tissue of an-

other plant so that the two will grow together. The inserted bud develops into a new shoot. *See* grafting.

■ buffer

a solution that resists change in pH when an acid or alkali is added, or when solutions are diluted.

■ buoyant density

the intrinsic density which a molecule, virus or sub-cellular particle has when suspended in an aqueous solution of a salt, such as CsCl, or a sugar, such as sucrose. DNA from different species has a characteristic buoyant density, which reflects the proportion of G=C base pairs. The greater the proportion of G=C, the greater the buoyant density of the DNA.

■ CAAT box

(also known as CAT box) a conserved sequence found within the promoter region of the protein-encoding genes of many eukaryotic organisms. It has the consensus sequence GGCCAATCT; it occurs around 75 bases prior to the transcription initiation site; and it is one of several sites for recognition and binding of regulatory proteins called transcription factors.

■ callus

1. a protective tissue, consisting of parenchyma cells, that develops over a cut or damaged plant surface.
2. mass of unorganised, thin-walled parenchyma cells induced by hormone treatment.
3. actively dividing non-organised masses of undifferentiated and differentiated cells often developing from injury (wounding) or in tissue culture in the presence of growth regulators.

■ callus culture

a technique of tissue culture; it is usually on solidified medium and initiated by inoculation of small explants or sections from established organ or other cultures (the inocula). Callus culture is used as the basis for organogenic (shoot, root) cultures, cell cultures or proliferation of embryoids. Callus cultures can be indefinitely maintained through regular sub-culturing.

■ **calorie**

equivalent to the amount of heat required to raise the temperature of 1 gram of water from 14.5°C to 15.5°C (»4.19 J). *cf* kilocalorie.

■ **calyx**

all the sepals of a flower considered collectively. The outermost whorl of flower parts.

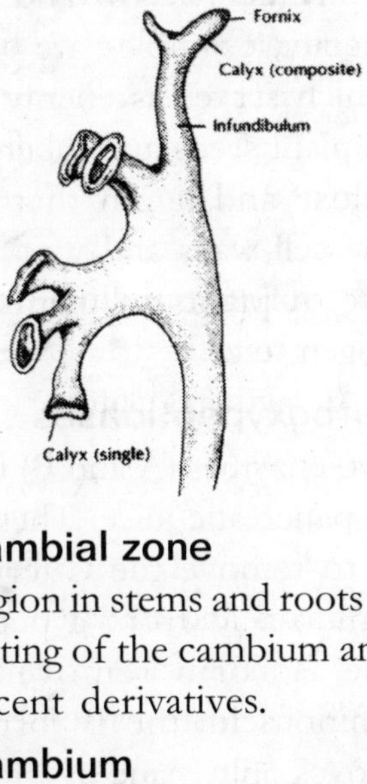

■ **cambial zone**

region in stems and roots consisting of the cambium and its recent derivatives.

■ **cambium**

(L. *cambium*, one of the alimentary body fluids supposed to nourish the body organs; pl: cambia). A layer, usually regarded as one or two cells thick, of persistently meristematic tissue between the xylem and phloem tissues, and which gives rise to secondary tissues, thus resulting in an increase in diameter. The two most important cambia are the vascular (fascicular) cambium and the cork cambium.

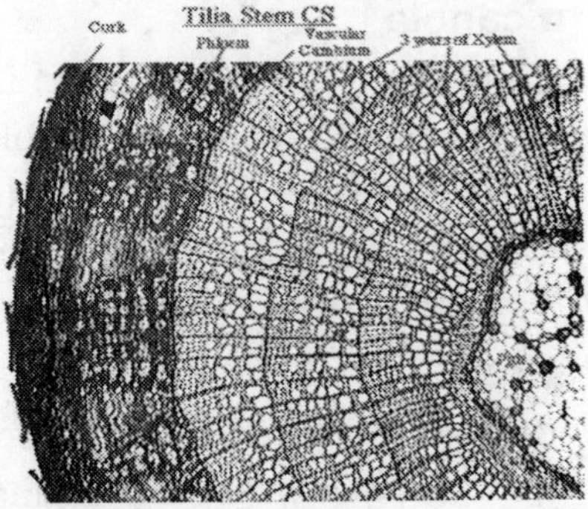

■ **cancer**

uncontrolled growth of the cells of a tissue or an organ in a multicellular organism. *cf* oncogenesis.

■ **candidate gene**

a gene whose function suggests that it may be involved in the genetic variation observed for a particular trait, e.g., the gene for growth hormone is a candidate gene for body weight.

■ **candidate-gene strategy**

an experimental approach in which knowledge of the biochemistry and/or physiology of a trait is used to draw up a list of genes whose protein products could be involved in the trait.

■ **canine**

pertaining to dogs.

■ **canola**

any of several cultivars of oil-seed rape (more fully: canola oil); the vegetable oil high in mono-unsaturated fatty acid obtained from these cultivars.

■ **cap**

the structure found on the 5´-end of eukaryotic mRNA, and consisting of an inverted, methylated guanosine residue.

■ **cap site**

the site in a gene where translation is initiated, a.k.a. translation initiation site.

■ **capacitation**

the final stage in the maturation process of a spermatozoon, taking place inside the female genital tract as the sperm penetrates the ovum.

■ **capsid**

the protein coat of a virus. The capsid often determines the shape of the virus.

■ **carbohydrate**

an organic compound based on the general formula $C_x(H_2O)_y$, performing many vital roles in living organisms. The simplest carbohydrates are the sugars (saccharides), including glucose and sucrose. Polysaccharides are carbohydrates of much greater molecular weight and complexity; examples are starch, which serves as energy store in plant seeds and tubers; cellulose and lignin that form the cell walls and woody tissue of plants of plants; glycogen, etc.

■ **carboxypeptidases**

two enzymes (A and B) found in pancreatic juice. Their role is to remove the C-terminal amino-acid from a peptide; the A form removes any amino acid; the B form removes only lysine or arginine. Used when sequencing peptides.

■ **carcinogen**

a substance capable of inducing cancer in an organism.

■ carcinoma

a malignant tumour derived from epithelial tissue, which forms the skin and the outer cell layers of internal organs.

■ carotene

a reddish-orange plastid pigment involved in light reactions in photosynthesis.

■ carotenoid

red to yellow pigments responsible for the characteristic colour of many plant organs or fruits, such as tomatoes, carrots, etc. Oxidation products of carotene are called xanthophylls. Carotenoids serve as light-harvesting molecules in photosynthetic assemblies and also play a role in protecting prokaryotes from the deleterious effects of light.

■ carpel

female reproductive organ of flowering plants, consisting of stigma, style and ovary. In some plants, one or more carpels unite to form the pistil.

■ carrier

in genetics, typically an individual that has one recessive mutant allele for some defective condition that is "masked" by a dominant normal allele at the same locus, i.e., an individual that is heterozygous for a recessive harmful allele and a dominant normal allele; the phenotype is normal, but the individual passes the defective (recessive) allele to half of its offspring.

■ carrier DNA

DNA of undefined sequence content which is added to the transforming (plasmid) DNA used in physical DNA-transfer procedures. This additional DNA increases the efficiency of transformation in electroporation and chemically mediated DNA-delivery systems. The mechanism responsible for this effect is not known. *See also* binary vector; plasmid; chimeric gene.

■ carrier gas

the gas that carries the sample in gas chromatography.

■ casein

a group of proteins found in milk.

casein hydrolysate
mixture of amino acids and peptides produced by enzymatic or acid hydrolysis of casein. *cf* organic complex; undefined.

catabolic pathway
a pathway by which an organic molecule is degraded in order to release energy for growth and other cellular processes; degradative pathway.

catabolism
the metabolic breakdown of large molecules in living organism, with accompanying release of energy.

catabolite repression
glucose-mediated reduction in the rates of transcription of operons that encode enzymes involved in catabolic pathways (such as the *lac* operon).

catalysis
the process of changing the rate of a chemical reaction by use of a catalyst.

catalyst
a substance that promotes a chemical reaction by lowering the activation energy of a chemical reaction, without itself undergoing any permanent chemical change. The process is catalysis. *cf* catalytic antibody; catalytic RNA.

catalytic antibody
an antibody selected for its ability to catalyse a chemical reaction by binding to and stabilising the transition-state intermediate.

cDNA
the double-stranded DNA complement of an mRNA sequence; synthesised *in vitro* from a mature RNA template using reverse transcriptase (to create a single strand of DNA from the RNA template) and DNA polymerase (to create the double-stranded DNA). Preparation of cDNAs is often the first step in cloning DNA sequences of interest. Used as specific and sensitive probes in hybridisation studies, because cDNAs usually do not include regulatory or other controlling sequences, and so they can be used to identify (probe) and isolate genes and their associated sequences from genomic DNA.

■ cDNA clone

a double-stranded DNA molecule that is carried in a vector and was synthesised *in vitro* from an mRNA sequence by using reverse transcriptase and DNA polymerase.

■ cDNA cloning

a method of cloning the coding sequence of a gene, starting with its mRNA transcript. It is normally used to clone a DNA copy of a eukaryotic mRNA. The cDNA copy, being a copy of a mature messenger molecule, will not contain any intron sequences and may be readily expressed in any host organism if attached to a suitable promoter sequence within the cloning vector.

cDNA library

a collection of cDNA clones that were generated *in vitro* from the mRNA sequences isolated from an organism or a specific tissue or cell type or population of an organism. *cf* library .

CDR

complementarity-determining regions,these are regions of the variable (V) regions of light and heavy antibody chains that make contact with the antigen. The primary amino acid sequences of these regions are highly variable among antibodies of the same class.

■ cell differentiation

continuous loss of physiological and cytological characters of young cells, resulting in getting the characters of adult cells. The unspecialised cells become modified and specialised for the performance of specific functions. Differentiation results from the controlled activation and de-activation of genes.

■ cell division

formation of two or more daughter cells from a single mother cell. The nucleus divides first, followed by the formation of a cell membrane between the daughter nuclei. Division of cytoplasm and nucleus into two or more parts by formation of a cell plate.

■ cell fusion

formation of a single hybrid cell from two cells of different species, cultured *in vitro*. The cells fuse and coalesce, but their nuclei may remain separated. During subsequent cell division, a single spindle is formed so that each daughter cell has a single nucleus containing sets of chromosomes from each parental line. Subsequent divisions often result in the loss of chromosomes and therefore of genes. The cell fusion technique can be used to determine the control of specific genes and their assignment to chromosomes. *cf* cell hybridisation.

■ **cell generation time**

the interval between the beginning of consecutive divisions of a cell. The time that it takes for a population of single-celled organisms to double its cell number. Successive generations of cells or organisms within a population are separated by a time interval called generation time. The cell regeneration time can be determined with the aid of time-lapse microcinematography.

■ **cell hybridisation**

the fusion of two or more dissimilar cells leading to the formation of a somatic hybrid. *cf* cell fusion.

■ **cell line**

a cell lineage that can be maintained in culture. A cell line arises from a primary culture. It implies that cultures from it consist of several lineages of the cells originally present in the primary culture.

■ **cell membrane**

the membrane that separates the cell wall and the cytoplasm, and regulates the flow of material into and out of the cell. *See* plasmalemma.

■ **cell number**

the number of cells per unit volume of a culture.

■ **cell plate**

the precursor of the cell wall, formed as cytokinesis starts during cell division. The cell plate develops in the region of the equatorial plate and arises from membranes in the cytoplasm.

■ **cell sap**

water and dissolved substances, sugar, amino acids,

waste substances, etc., in the plant cell vacuole.

■ **cell selection**

the process of selecting cells within a group of genetically different cells. Select cells or cell lines are sub-cultured onto fresh medium for continued selection and often are exposed to an increased level of the selection agent. The final objective is to regenerate plants exhibiting the traits selected for at the cellular level.

■ **cell strain**

a strain of cells having specific properties or markers derived from a primary culture of a cell line by selection or cloning. The selected properties must persist during subsequent cultivation. a.k.a. single-cell line.

■ **cell suspension**

cells in culture in moving or shaking liquid medium, often used to describe suspension cultures of single cells and cell aggregates. *See* suspension culture.

■ **cell wall**

a rigid external coat which surrounds plant cells. It is formed outside the plasmalemma and consists primarily of cellulose.

■ **cell-mediated immune response**

the activation of T cells of the immune system in response to the presence of a foreign antigen.

■ **cellular oncogene**

a normal mammalian or avian gene that when mutated or improperly expressed contributes to the development of cancer.

■ **cellulase**

enzyme catalysing the breakdown of cellulose.

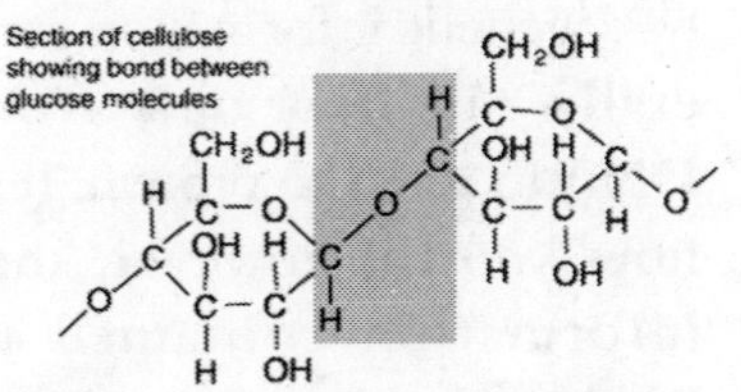

■ **cellulose**

a complex carbohydrate composed of long, unbranched chains of beta-glucose ((1.4)-linked-b-d-glucose) molecules, which contribute to the structural framework of plant cell walls. It comprises 40% to 55% by weight of the plant cell wall.

■ cellulose nitrate

see **nitrocellulose.**

■ cellulosome

a multi-protein aggregate that is present in some cellulolytic micro-organisms and contains multiple copies of all the enzymes required to completely break down cellulose. This complex is often found on the outer surface of cellulolytic micro-organisms.

■ centiMorgan (cM)

one percent recombination between two loci.

■ central dogma

the basic concept that, in nature, genetic information generally can flow only from DNA to RNA to protein. It is now known, however, that information contained in RNA molecules of certain viruses (called retroviruses) can also flow back to DNA.

■ central mother cell

a subsurface cell located in a plant apical meristem and characterised by a large vacuole.

■ centres of origin

the locations in the world where particular domesticated plants originated. These areas show the highest variation, and are rich in wild alleles.

■ centrifugation

separating molecules by size or density using centrifugal forces generated by a spinning rotor. G-forces of several hundred thousand times gravity are generated in ultracentrifugation.

■ centrifuge

a device in which solid or liquid particles of different densities are separated by rotating them in a tube in a horizontal circle. The denser particles tend to move along the length of the tube to a greater radius of rotation, displacing the lighter particles to the other end.

■ centriole

an organelle in many animal cells that appears to be involved in the formation of the spindle during mitosis. During cell division, the two centrioles move to opposite sides of the nucleus to form the ends of the spindle.

■ **centromere**
the portion of the chromosome to which the spindle fibres attach during mitotic and meiotic division. It appears as a constriction when chromosomes contract during cell division. After chromosomal duplication, which occurs at the beginning of every mitotic and meiotic division, the two resultant chromatids are joined at the centromere.

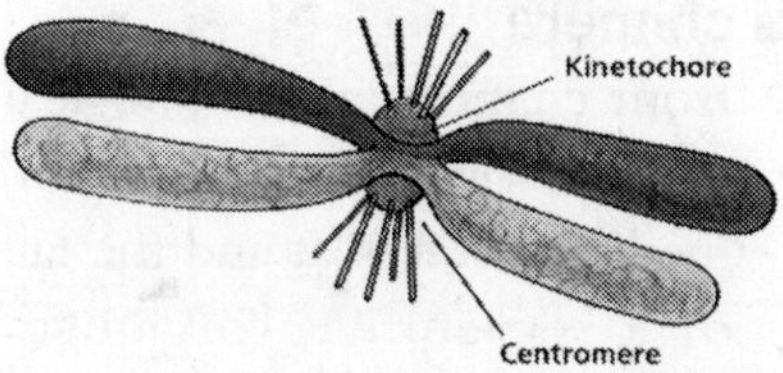

■ **centrosome**
a specialised region of a living cell, situated next to the nucleus, where micro-tubules are assembled and broken down during cell division. The centrosome of most animal cells contains a pair of centrioles. During metaphase the centrosome separates into two regions, each containing one of the centrioles.

■ **cephem-type antibiotic**
an antibiotic that shares the basic chemical structure of cephalosporin.

■ **chain terminator**
1. codons which do not code for an amino acid. They signal ribosomes to terminate protein synthesis. The codons are UAA, UAG and UGA, and have been termed ochre, amber and opal, respectively. Also known as stop codons or termination codons. Often two of these codons are found together at the end of a coding sequence of RNA.
2. in the Sanger method of DNA sequencing, dideoxynucleoside triphosphates are added as chain terminators in the synthesis of a complementary DNA strand.

■ **character**
a distinctive feature of an organism.

■ **chelate**
noun: complex organic molecule that can combine with cations and does not ionize. Chelates can supply micronu-

trients to plants at slow, steady rates. Usually used to supply iron to plant cells.

chemical mutagen

a chemical or product capable of causing genetic mutation in living organisms exposed to it.

chemically-defined medium

when all of the chemical components of a plant tissue culture medium are fully known and defined. *cf* undefined; organic complex.

chemiluminescence

the emission of light from a chemical reaction.

chemostat

a continuous and open culture in which growth rate and cell density are maintained constant by a fixed rate of input of a growth-limiting nutrient. *cf* phytostat.

chemotaxis

motion of a motile cell, organism or part towards or away from an increasing concentration of a particular substance.

chemotherapy

the treatment of disease, especially infections or cancer, by means of chemicals. In treating cancers, it involves administering chemicals toxic to malignant cells.

chiasma

two lines placed crosswise ,a visible point of junction between two non-sister chromatids of homologous chromosomes during the first meiotic prophase. a.k.a. cross-over. In the diplotene stage of prophase I of meiosis, the four chromatids of a bivalent are associated in pairs, but in such a way that one part of two chromatids is exchanged.

chimera

from chimera, a mythological creature with the head of a lion, the body of a goat and the tail of a serpent. An organism whose cells are not all derived from the same zygote.

1. animal. an individual exhibiting two or more genotypes in patches derived from two or more embryos. An individual derived from two embryos by experimental intervention.

2. plant. part of a plant with a genetically different constitution as compared with other

parts of the same plant. It may result from different zygotes that grow together, or from artificial fusion (grafting); it may either be periclinal chimera, in which one tissue lies over another as a glove fits a hand; mericlinal chimera, where the outer tissue does not completely cover the inner tissue; and sectoral chimera, in which the tissues lie side by side.

3. a recombinant DNA molecule that contains sequences from different organisms.

■ chimeric DNA

a recombinant DNA molecule containing unrelated genes.

■ chimeric gene

a semi-synthetic gene, consisting of the coding sequence from one organism, fused to promoter and other sequences derived from a different gene. Most genes used in transformation are chimeric. *See* carrier DNA; binary vector; plasmid; transformation; vector.

■ chimeric selectable marker gene

a gene that is constructed from parts of two or more different genes and allows the host cell to survive under conditions where it would otherwise die.

■ chitin

a nitrogenous polysaccharide occurring as skeletal material in many invertebrates and fungi.

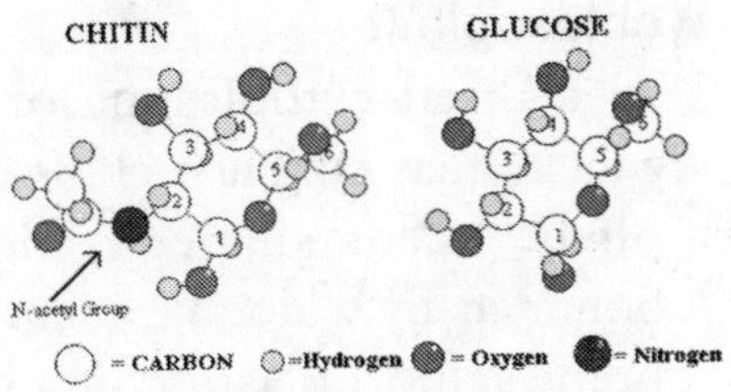

■ chitinase

an enzyme which breaks down chitin.

■ chloramphenicol

an antibiotic that interferes with protein synthesis.

■ chlorenchyma

tissue containing chloroplasts, including leaf mesophyll and other parenchyma cells.

■ chlorophyll

one of the two pigments responsible for the green colour of most plants. It is essential in the absorption of light energy for photosynthesis.

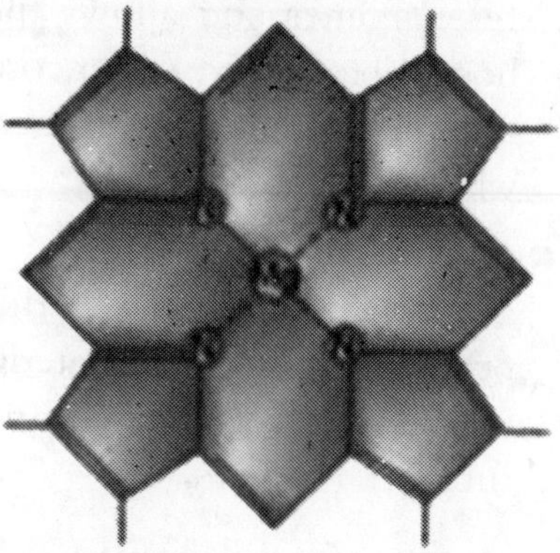

■ **chloroplast**

specialised cytoplasmic organelle that contains chlorophyll. Lens-shaped and bounded by a double membrane, chloroplasts contain membranous structures (thylakoids) piled up into stacks, surrounded by a gel-like matrix (stroma). They are the site of solar energy transfer and important reactions of starch or sugar synthesis. Chloroplasts have their own DNA and are inherited cytoplasmically, independent of nuclear genes.

■ **chlorosis**

failure of chlorophyll development, and appearance of yellow colour in plants, because of a nutritional disturbance or because of an infection by a virus, bacteria or fungus.

■ **chromatid**

each of the two daughter strands comprising a duplicated chromosome. The term remains in use while the two chromatids are still joined at the centromere. As soon as the centromere divides, setting the two chromatids adrift (during anaphase of mitosis; and during anaphase II of meiosis), they are called chromosomes.

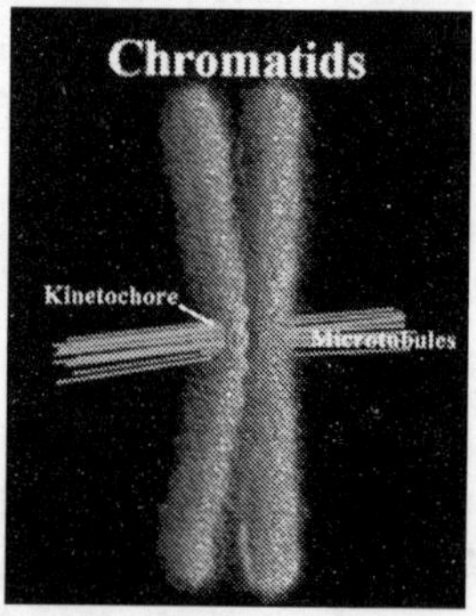

■ **chromatin**

substance of which eukaryotic chromosomes are composed. It consists of primarily DNAm with some proteins (mainly histones), and small amounts of RNA. Originally named because of the readiness with which it stains with certain dyes (chromaticity).

■ **chromatin fibres**

a basic organisational unit of eukaryotic chromosomes, consisting of DNA and associated proteins assembled into strands of 30 nm average diameter.

■ **chromatography**

1. a method for separating and identifying the components of mixtures of molecules having similar chemical and physical properties.
2. the term originally used by Mikhail Tswett (1906) to describe the separation of a mixture of leaf pigments on a calcium carbonate column.

■ **chromocentre**

body produced by fusion of the heterochromatic regions of the chromosomes in the polytene tissues (e.g., the salivary glands) of certain *Diptera*.

■ **chromogenic substrate**

a compound or substance that contains a colour-forming group.

■ **chromomeres**

small bodies, described by J. Belling, that are identified by their characteristic size and linear arrangement along a chromosome.

■ **chromonema**

sn optically single thread forming an axial structure within each chromosome.

■ **chromoplast**

plastid containing pigments, such a chloroplast, or one in which carotenoids predominate.

■ **chromosomal aberration**

any change in chromosome structure or number. Although it can be a mechanism for enhancing genetic diversity, such alterations are usually fatal or ill-adaptive, especially in animals.

■ **chromosomal integration site**

a chromosomal location where foreign DNA can be integrated, often without impairing any essential function in the host organism.

■ **chromosomal polymorphism**

the occurrence of one to several chromosomes in two or more alternative structural forms within a population; the structurally changed chromosomes are the result of chromosome mutations (i.e., any structural change involving the gain, loss or re-location of chromosome segments).

■ **chromosome**

1.a single DNA molecule, a tightly coiled strand of DNA, condensed into a compact structure *in vivo* by complexing with accessory histones or histone-like proteins.

2.a group of nuclear bodies containing genes which are largely responsible for the differentiation and activity of a eukaryotic cell; one of the bodies into which the nucleus resolves itself at the beginning of mitosis and from which it is derived at the end of mitosis. Chromosomes contain most of the cell's DNA. Chromosomes exist in pairs in eukaryotes - one paternal (from the male parent) and one maternal (from the female parent). Each eukaryotic species has a characteristic number of chromosomes. Bacterial and viral cells contain only a single chromosome, consisting of a single or double strand of DNA or, in some viruses, RNA, without histones.

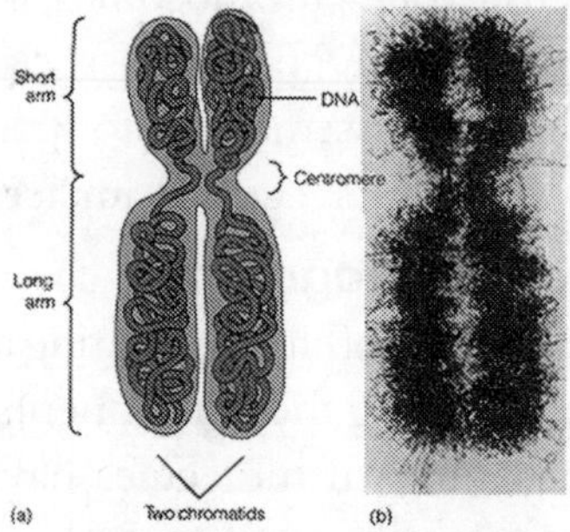

■ **chromosome aberration**

abnormal structure or number of chromosomes; includes deficiency, duplication, inversion, translocation, aneuploidy, polyploidy, or any other change from the normal pattern.

■ **chromosome banding**

staining of chromosomes in such a way that light and dark areas occur along the length of the chromosomes in repeatable patterns. Lateral comparisons identify pairs. Each chromosome can be identified by its banding pattern.

■ **chromosome jumping**

a technique that allows two segments of duplex DNA that are separated by thousands of base pairs (about 200 kb) to be cloned together. After subcloning, each segment can be used as a probe to identify cloned DNA sequences that, at the chromosome level, are roughly 200 kb apart..

■ chromosome mutation

a change in the gross structure of a chromosome, usually causing severely deleterious effects in the organism. They are often due to an error in pairing during the crossing-over stage of meiosis. The main types of chromosome mutation are translocation, duplication, deletion and inversion.

■ chromosome theory of inheritance

the theory that chromosomes carry the genetic information and that their behaviour during meiosis provides the physical basis for segregation and independent assortment.

■ chromosome walking

a technique that identifies overlapping cloned DNA fragments that form one continuous segment of a chromosome. These fragments can be generated either by random shearing or by partial digestion with a four-base-pair cutter such as *Sau*3A. A series of colony hybridisations is then carried out, starting with some cloned fragment which has already been identified and which is known to be in the region encompassed by the overlapping clones. This identified fragment is used as a probe to pick out clones containing adjacent sequences. These are then used as probes themselves to identify clones carrying sequences adjacent to them and so on. At each round of hybridisation one 'walks' further along the chromosome from the initial fragment.

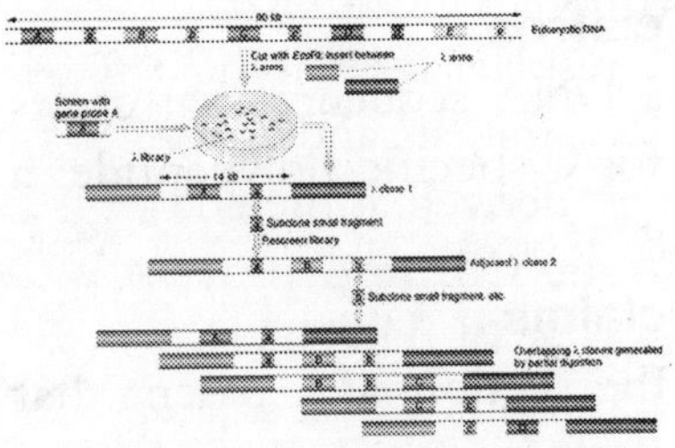

■ chymosin

an enzyme that clots milk; it is used in the manufacture of cheese.

■ circularisation

a DNA fragment generated by digestion with a single restriction endonuclease will have complementary 5′ and 3′ extensions (sticky ends). If these ends are annealed and ligated, the DNA fragment will have been converted to a covalently-closed circle, or circularised.

■ **cis heterozygote**
a heterozygote that contains two mutations arranged in a cis configuration (e.g., *a*+ *b*+ / *a b)*.

■ **cis-acting sequence**
a nucleotide sequence that only affects the expression of genes located on the same chromosome.

■ **cistron**
a DNA sequence that codes for a specific polypeptide; a gene.

■ **claims**
the section of a patent that states, in detail, the uses and possible applications of the invention described in the patent.

■ **class switching**
the process during which a plasma cell stops producing antibodies of one class and begins producing antibodies of another class.

■ **cleave**
to break phospho-diester bonds of double-stranded DNA, usually with a type II restriction endonuclease. a.k.a. to cut or digest.

■ **clonal propagation**
asexual propagation of many new plants (ramets) from an individual (ortet); all have the same genotype.

■ **clonal selection**
the production of a population of plasma cells all producing the same antibody in response to the interaction between a B lymphocyte producing that specific antibody and the antigen bound by that antibody.

■ **clone**
1. a group of cells or organisms that are genetically identical as a result of asexual reproduction, breeding of completely inbred organisms, or forming genetically identical organisms by nuclear transplantation.
2. group of plants genetically identical in which all are de-

rived from one selected individual by vegetative propagation, without the sexual process.

3. a population of cells that all carry a cloning vehicle with the same insert DNA molecule.

4. verb: To clone. To insert a DNA segment into a vector or host chromosome. *See* cloning.

■ cloned strain or line

a strain or line descended directly from a clone.

■ cloning

1. the mitotic division of a progenitor cell to give rise to a population of identical daughter cells or clones.

2. incorporation of a DNA molecule into a chromosomal site or a cloning vector.

3. animal cloning: the creation of a whole animal by mitotic divisions from a single diploid somatic cell , typically by the process of nuclear transfer (q.v.). Cloning by nuclear transfer from undifferentiated embryonic cells has been possible for many years, but its widespread application has been hampered by inability to culture embryonic cells from animals other than mice. In 1997, Ian Wilmut and colleagues from Edinburgh showed that it is possible to create a whole animal from a cell taken from differentiated adult tissue, thereby opening up the possibility of widespread animal cloning.

■ cloning vector

a small, self-replicating DNA molecule - usually a plasmid or viral DNA chromosome - into which foreign DNA is inserted in the process of cloning genes or other DNA sequences of interest. It can carry inserted DNA and be perpetuated in a host cell. Also called a cloning vehicle, vector, or vehicle.

■ closed continuous culture

a continuous culture in which inflow of fresh medium is balanced by outflow of corresponding volumes of spent medium. Cells are separated mechanically from outflowing medium and added back to the culture. *cf* open continuous culture; batch culture; continuous culture.

■ **coccus**

a spherical bacterium. Cocci may occur singly, in pairs, in groups of four or more, and in cubical packets.

■ **coconut milk**

liquid endosperm of the coconut, often used to supply organic nutrients to cultured cells and tissues. *See* addendum; organic complex; undefined.

■ **cocoon**

a protective coverage for eggs and/or larvae produced by many invertebrates, such as the silkworm moth.

■ **co-culture**

the joint culture of two or more types of cells, such as a plant cell and a micro-organism, or two types of plant cells. Used in various dual-culture systems or in nurse-culture.

■ **coding**

the specification of a peptide sequence, by the code contained in DNA molecules.

■ **coding sequence**

that portion of a gene which directly specifies the amino acid sequence of its protein product. Non-coding sequences of genes include control regions, such as promoters, operators and terminators, as well as the intron sequences of certain eukaryotic genes.

■ **coding strand**

the strand of duplex DNA which contains the same base sequence (after substituting U for T) found in the mRNA molecule resulting from transcription of that segment of DNA. a.k.a. sense strand. The mRNA molecule is transcribed from the other strand, known as the template or antisense strand.

Coding strand 5´ ATGAAAGCTTTAGTGGGCGCCCGTAT 3´

template strand 3´ TACTTTCGAAATCACCC GCGGGCATA 5´
mRNA 5´ AUGAAAGCUU UAGUGGGCGCCCGUAU 3´

■ **co-dominance**

the situation in which both alleles in a heterozygous individual are expressed, so that the phenotype of heterozygotes incorporates the phenotypic effect of each allele. For example, roan coat colour in cattle results from a mixture of red hairs and white hairs, caused by heterozygosity for the red allele and the white allele. Also, protein polymorphisms and microsatellites show co-dominance: heterozygotes have two bands, whereas homozygotes have only one band.

■ **co-dominant alleles**

alleles that produce independent effects when in the heterozygous condition.

■ **codon**

a set of three nucleotides in mRNA, functioning as a unit of genetic coding by specifying a particular amino acid during the synthesis of polypeptides in a cell. A codon specifies a transfer RNA carrying a specific amino acid, which is incorporated into a polypeptide chain during protein synthesis. The specificity for translating genetic information from DNA into mRNA, then to protein, is provided by codon-anticodon pairing.

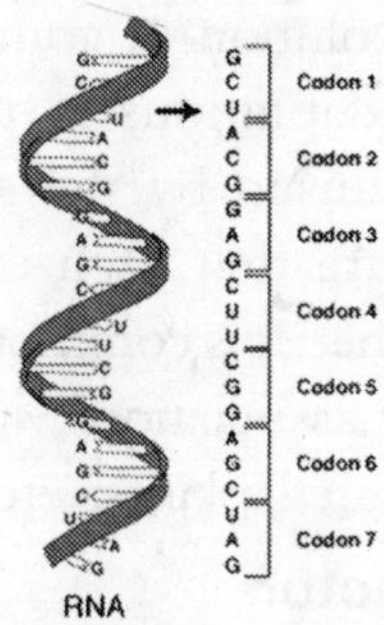

Ribonucleic acid

■ **codon optimisation**

an experimental strategy in which codons within a cloned gene - ones not generally used by the host cell translation system - a re changed by *in vitro* mutagenesis to the preferred codons, without changing the amino acids of the synthesised protein.

■ **coefficient**

a number expressing the amount of some change or

effect under certain conditions (e.g., the coefficient of inbreeding).

■ **co-enzyme**
an organic molecule of low molecular weight and usually non-protein, such as a vitamin, that binds to an enzyme and promotes its catalytic activity.

■ **co-evolution**
the evolution of complementary adaptations in two species caused by the selection pressure that each exerts on the other. It is common in symbiotic associations, in insect-pollinated plants, etc.

■ **co-factor**
an organic molecule or inorganic ion necessary for the normal catalytic activity of an enzyme.

■ **co-fermentation**
the simultaneous growth of two micro-organisms in one bioreaction.

■ **co-generation**
production of both electricity and process heat (steam) in an industrial plant.

■ **cohesion**
holding together; a force holding a solid to a solid or a solid to a liquid, owing to attraction between like molecules.

■ **cohesive ends**
double-stranded DNA molecules with single-stranded ends which are complementary to each other, enabling the different molecules to join each other. a.k.a. protruding ends; sticky ends; overhang. *cf* cos ends.

■ **coincidence**
the ratio of the observed frequency of double crossings-over to the expected frequency, where the expected frequency is calculated by assuming that the two crossing-over events occur independently of each other.

■ **co-integrate**
a DNA molecule formed by the fusion of two different DNA molecules, usually mediated by a transposable element.

■ **co-integrate vector system**
a two-plasmid system for transferring cloned genes to

plant cells. The cloning vector has a T-DNA segment that contains cloned genes. After introduction into *Agrobacterium tumiefasciens*, the cloning vector DNA undergoes homologous recombination with a resident disarmed Ti plasmid to form a single plasmid carrying the genetic information for transferring the genetically engineered T-DNA region to plant cells.

■ colchicine

an alkaloid obtained from *Colchicum autumnale*, autumn crocus or meadow saffron, which inhibits spindle formation in cells during mitosis, so that chromosomes cannot separate during anaphase, thus inducing multiple sets of chromosomes. Also used to halt mitosis at metaphase - the stage when chromosomes are most visible.

■ coleoptile

protective sheath covering the shoot apex of the embryo in monocotyledenous plants. It protects the plumule as it emerges through the soil.

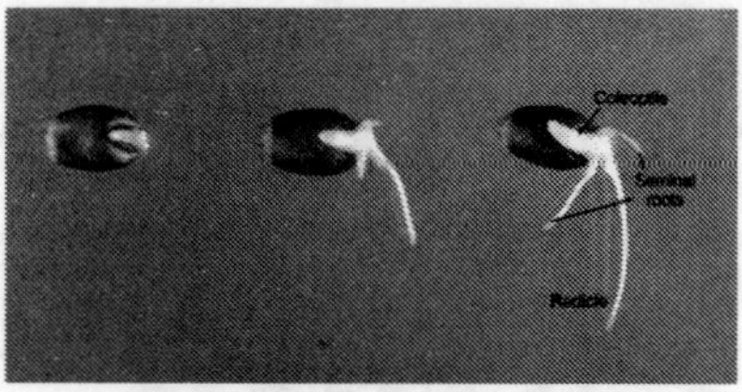

■ coleorhiza

a protective sheath surrounding the radicle of monocotyledenous plants.

■ co-linearity

a relationship in which the units in one molecule occur in the same sequence as the units in another molecule which they specify; e.g., the nucleotides in a gene are co-linear with the amino acids in the polypeptide encoded by that gene.

■ collenchyma

derived from parenchyma and denoting a type of cell tissue A tissue of living cells, the walls being unevenly thickened with cellulose and hemicellulose, but never lignified; it functions in mechanical support in young, short-lived or non-woody organs and is thus found in midribs and leaf petioles.

■ **colony**

1. an aggregate of identical cells (clones) derived from a single progenitor cell.
2. a group of interdependent cells or organisms.

■ **colony hybridisation**

a technique that uses a nucleic acid probe to identify a bacterial colony with a vector carrying a specific cloned gene or genes.

■ **combinatorial library**

during the ligation reaction with cDNAs of light and heavy antibody chains into a bacteriophage lambda (1) vector, many novel combinations consisting of one heavy and one light chain coding region are formed. The library comprises these combinations, each in a separate vector.

■ **commensalism**

the interaction of two or more dissimilar organisms where the association is advantageous to one without affecting the other(s). *cf* parasitism; symbiosis.

■ **companion cell**

living cell associated with the sieve cell of phloem tissue in vascular plants.

■ **comparative gene mapping**

the comparison of map locations of genes between species. The results of these comparisons indicate substantial conservation of blocks of genes and even large segments of chromosomes between species. Great use can be made of this conservation of map position. For example, in the case of mammals, it means that if a gene has been mapped in one or both of the intensely-mapped species (humans and mice), then the likely location of that gene in other mammals can be predicted with considerable confidence. Conversely, if a mapped anonymous DNA marker has an effect on a quantitative trait (this being indicative of the marker being linked to a quantitative trait locus (QTL)) in, say, cattle, then knowledge of the comparative map between cattle and humans can identify genes in the homologous region of the human genome that could correspond to the QTL. Such genes are called comparative positional candidate genes (q.v.).

■ **comparative positional candidate gene**
a gene that is likely to be located in the same region as a DNA marker that has been shown to be linked to a single-locus trait or to a quantitative trait locus (QTL), where the gene's likely location in the genome of the species in question is based on its known location in the map of another species, i.e., is based on the comparative map between the two species.

■ **competence**
ability of a bacterial cell to take up DNA molecules and become genetically transformed.

■ **competency**
an ephemeral state, induced by treatment with cold cations, during which bacterial cells are capable of taking up foreign DNA.

■ **competent**
a competent cell is capable of developing into a fully functional embryo. The opposite is non-competent.

■ **complement proteins**
proteins that bind to antibody-antigen complexes and help degrade the complexes by proteolysis.

■ **complementarity**
the relationship between the two strands of a double helix of DNA. Thymine in one strand pairs with adenine in the other strand, and cytosine in one strand pairs with guanine in the other strand.

■ **complementary entity**
1. one of a pair of nucleotide bases that form hydrogen bonds with each other. Adenine (A) pairs with thymine (T) [or with Uracil (U) in RNA], and guanine (G) pairs with cytosine (C).

2. one of a pair of segments or strands of nucleic acid that will hydridise (join by hydrogen bonding) with each other.

■ **complementary genes**

two or more interdependent genes, such that (in the case of dominant complementarity) the dominant allele from either gene can only produce an effect on the phenotype of an organism if the dominant allele from the other gene is also present; or (in the case of recessive complementarity) only double homozygous recessive show the effect.

■ **complementary homopolymeric tailing**

the process of adding complementary nucleotide extensions to different DNA molecules, e.g., dG (deoxyguanosine) to the 3´-hydroxyl ends of one DNA molecule and dC (deoxycytidine) to the 3´-hydroxyl ends of another DNA molecule to facilitate, after mixing, the joining of the two DNA molecules by base pairing between the complementary extensions. Also called dG - dC tailing, dA - dT tailing.

■ **complementary nucleotides**

members of the pairs adenine-thymine, adenine-uracil, and guanine-cytosine that have the ability to hydrogen bond to one another.

■ **complementation test**

introduction of two mutant chromosomes into the same cell to determine whether the mutants are alleles of the same gene. If the mutations are non-allelic, the genotype will be m_1 +/+ M_2 , and the phenotype will be wild-type (normal), because each chromosome "covers" for the other. In contrast, if they are allelic, the mutant phenotype will result.

■ **complete digest**

the treatment of a DNA preparation with an endonuclease for sufficient time for all of the potential target sites within that DNA to have been cleaved. *cf* partial digest.

■ **composite transposon**

a transposable element formed when two identical or nearly identical transposons insert on either side of a non-transposable segment of DNA, such as the bacterial transposon Tn5.

Composite transposons

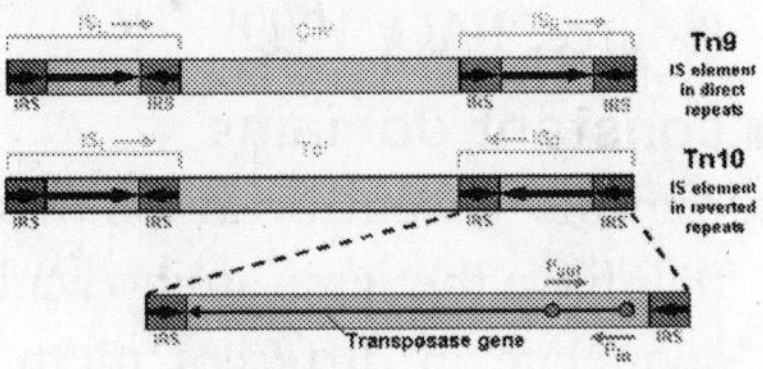

■ **compound chromosome**

a chromosome formed by the union of two separate chromosomes, as in attached-X chromosomes or attached-X-Y chromosomes.

■ **concatemer**

a DNA segment made up of repeated sequences linked end to end.

■ **concordance**

identity of matched pairs or groups for a given trait, such as sibs expressing the same trait.

■ **conditional lethal mutation**

a mutation that is lethal under one set of environmental conditions - the restrictive conditions - but is viable under another set of environmental conditions - the permissive conditions, e.g., temperature-sensitive mutations.

■ **conditioning**

1. the effects on phenotypic characters of external agents during critical developmental stages.

2. the undefined interaction between tissues and culture medium resulting in the growth of single cells or small aggregates. Conditioning may be accomplished by immersing cells or callus contained within a porous material (such as dialysis tubing) into fresh medium for a period dependent on cell density and a volume related to the amount of fresh medium.

■ **conidium**

an asexual spore produced by a specialised hypha in certain fungi.

■ **conjugation**

1. union of sex cells (gametes) or unicellular organisms during fertilisation.

2. the unidirectional transfer of DNA (bacterial plasmid) from one bacterium cell to another and involving cell-to-

cell contact. The plasmid usually encodes the majority of the functions necessary for its own transfer.

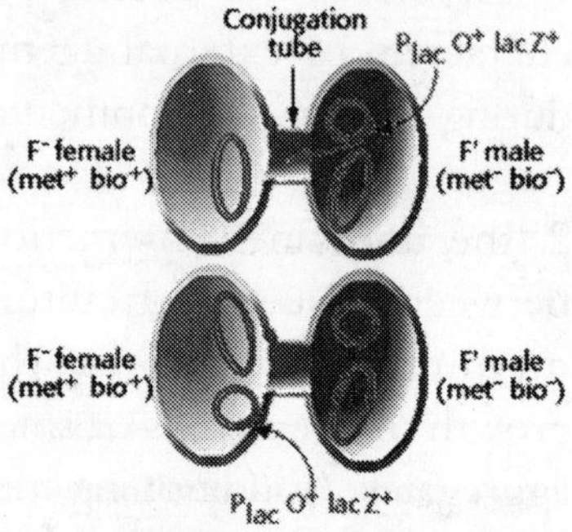

■ **conjugative functions**
plasmid-based genes and their products that facilitate the transfer of a plasmid from one bacterium to another.

■ **consanguinity**
related by descent from a common ancestor.

■ **consensus sequence**
the nucleotide sequence that is present in the majority of genetic signals or elements that perform a specific function.

■ **conservation of farm animal genetic resources**
In AnGR: Refers to all human activities, including strategies, plans, policies and actions, undertaken to ensure that the diversity of farm animal genetic resources is being maintained to contribute to food and agricultural production and productivity; now and in the future. (Source: FAO, 1999).

■ **constant domains**
regions of antibody chains that have the same amino acid sequence in different members of a particular class of antibody molecules.

■ **constitutive**
an organism is said to be constitutive for the production of an enzyme or other protein if that protein is always produced by the cells under all physiological conditions. *See* inducible.

■ **constitutive enzyme**
an enzyme that is synthesised continually regardless of growth conditions. *cf* inducible enzyme; repressible enzyme.

■ **constitutive gene**
a gene that is continually expressed in all cells of an organism.

■ **constitutive promoter**
an unregulated promoter that allows for continual transcription of its associated gene.

■ **constitutive synthesis**
continual production of RNA or protein by an organism.

■ **contaminant**
bacterial, fungal or algal micro-organism accidentally introduced into a culture or culture medium. Contaminant may overgrow the plant cells and consequently inhibit their growth. Working under aseptic conditions with a rigorous exclusion of potential contaminants must be practised in plant tissue culture.

■ **contig**
a set of overlapping clones that provide a physical map of a portion of a chromosome.

■ **contiguous map**
the alignment of sequence data from large, adjacent regions of the genome to produce a continuous nucleotide sequence across a chromosomal region..

■ **continuous culture**
a suspension culture continuously supplied with nutrients by the inflow of fresh medium. The culture volume is normally constant. *cf* batch culture; closed continuous culture; continuous fermentation; open continuous culture.

■ **continuous fermentation**
a process in which cells or micro-organisms are maintained in culture in the exponential growth phase by the continuous addition of fresh medium that is exactly balanced by the removal of cell suspension from the bioreactor. *cf* continuous culture.

■ **continuous variation**
variation not represented by distinct classes. Phenotypes grade into each other, and measurement data are required for analysis. Multiple genes are usually responsible for this type of variation. a.k.a. quantitative variation. *cf* discontinuous variation.

■ **control**
1. noun: unchanged (standard) protocol or treatment for comparison with the experimental treatment. The term is commonly used for untreated organisms.
2. verb: to direct or regulate cultures with addition of plant growth regulators.

controlled environment
the environment in which parameters, such as light, temperature, relative humidity and sometimes the partial gas pressure, are fully controlled.

controlling element
in eukaryotes, transposable elements which control the activity of standard genes. A controlling element may, in the simplest case, inhibit the activity of a gene through becoming integrated in, or close to, that gene. Occasionally, either in germinal or somatic tissue, it may be excised from this site, and due to excision the activity of the gene is more or less restored, while the element may become reintegrated elsewhere in the genome where it may affect the activity of another gene. For example, in maize, a controlling element such as *Ac* or *Ds* is capable of influencing the expression of a nearby gene.

conversion
the development of a somatic embryo into a plant.

coordinate repression
correlated regulation of the structural genes in an operon by a molecule that interacts with the operator sequence.

co-polymers
mixtures consisting of more than one monomer; for example, polymers of two kinds of organic bases, such as uracil and cytosine (poly-UC) have been combined for studies of the genetic code.

copy DNA
see **cDNA.**

copy number
the average number of molecules of a plasmid or gene per genome contained in a cell.

co-repressor
an effector molecule that forms a complex with a repressor and turns off the expression of a gene or set of genes.

corpus
the corpus is found below the tunica (q.v.) and is a part of the apical meristem. In the corpus, cells divide in all directions, giving them an in-

crease in volume. *See* apical meristem.

■ correlation

a statistical association between variables.

■ cortex

primary tissue of a stem or root, bounded externally by the epidermis and internally in the stem by the phloem and in the root by the pericycle.

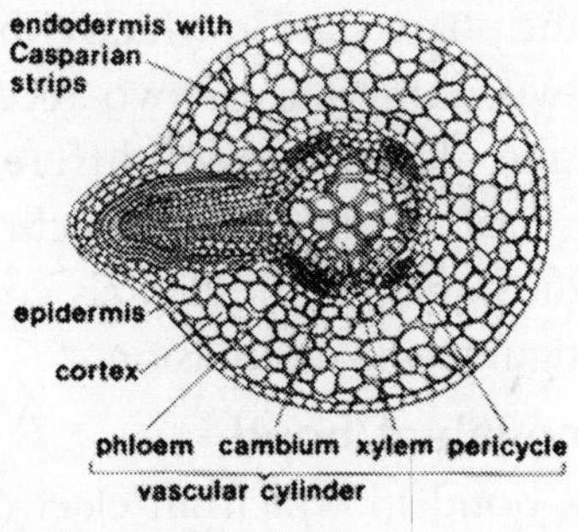

■ cos ends

the 12-base, single-strand, complementary extensions of bacteriophage lambda (l) DNA. a.k.a. cos sites.

■ co-segregation

when two genetic conditions appear to be inherited together.

■ cosmid

a plasmid vector which contains the two cos (cohesive) ends of phage lambda (λ) and one or more selectable markers such as an antibiotic resistance gene. Cosmids exploit certain properties of phage lambda (λ) to enable large, 40-50 kb, DNA fragments to be cloned at high efficiency. Cosmids and cosmid recombinants replicate as plasmids.

■ cot curve

when duplex DNA is heated, it dissociates into single strands. When the temperature is lowered, complementary strands tend to anneal or re-nature. The extent of re-naturation depends on the product of DNA concentration in moles of nucleotides per litre, and time in seconds. A graph showing the proportion of re-natured DNA against cot is known as a cot curve. The cot at which half the DNA has re-natured is the half-cot, a parameter indicating the degree of complexity of the DNA.

■ co-transfection

in baculovirus expression systems, the procedure by which the baculovirus and the transfer vector are simultaneously introduced into insect cells in culture.

■ co-transformation

in genetic engineering experiments, it is often necessary to transform with a plasmid for which there is no selectable phenotype and then screen for the presence of that plasmid within the host cell. Co-transformation is a technique in which host cells are incubated with two types of plasmid, one of which is selectable and the other not. Cells which have been transformed with the first plasmid are then selected. If transformation has been carried out at high DNA concentration, then it is probable that these cells will also have been transformed with the second (non-selectable) plasmid. The technique is frequently used in experiments with mammalian cells.

■ cotyledons

leaflike structures at the first node of the seedling stem. In some dicotyledons, they contain stored food for the young plant not yet able to photosynthesise its own food. Often referred to as seed leaves.

■ coupling

the phase state in which either two dominant or two recessive alleles of two different genes occur on the same chromosome. Also called cis configuration. *cf* repulsion.

■ covalent bond

a bond in which an electron pair is equally shared by protons in two adjacent atoms.

■ covalently closed circle

a double-stranded DNA molecule with no free ends. The two strands are interlinked and will remain together even after denaturation. In its native form, a CCC will adopt a supercoiled configuration.

■ co-variance

a measure of the statistical association between variables;

the extent to which two variables vary together.

■ cpDNA

the DNA of plant plastids, including chloroplasts.

■ critical breed

in AnGR: A breed where the total number of breeding females is less than 100 or the total number of breeding males is less than or equal to five; or the overall population size is close to, but slightly above 100 and decreasing, and the percentage of purebred females is below 80%.

■ critical-maintained breed

in AnGR: categories where critical or endangered breeds are being maintained by an active public conservation programme or within a commercial or research facility.

■ cross

in genetic studies, the mating of two individuals or populations. Also called mating.

■ cross hybridisation

the hydrogen bonding of a single-stranded DNA sequence that is partially but not entirely complementary to a single-stranded substrate. Often, this involves hybridising a DNA probe for a specific DNA sequence to the homologous sequences of different species.

■ cross pollination

fertilisation of a plant from a plant with a different genetic makeup.

■ cross pollination efficiency

efficiency of pollen from one plant reaching the stigma of another plant.

■ cross-breeding

mating between members of different populations (lines, breeds, races or species).

■ crossing over

a process in which homologous chromosomes exchange material through the breakage and reunion of two chromatids. A single crossover represents one reciprocal breakage and reunion event. A double crossover requires two simultaneous reciprocal breakage and reunion events. a.k.a. recombination; recombination event.

■ crossing-over unit

a measure of distance between two loci on genetic maps that is based on the average number of crossing-over events that take place in the interval between those two loci during meiosis. A map interval that is one crossing-over unit in length (a centiMorgan) describes an interval between two loci such that one in every hundred gametes recovered from meiosis is recombinant in that interval, i.e., the allele at the first locus is maternal in origin, while the allele at the other locus in that same gamete is paternal in origin.

■ crossover

a recombinant chromosome.

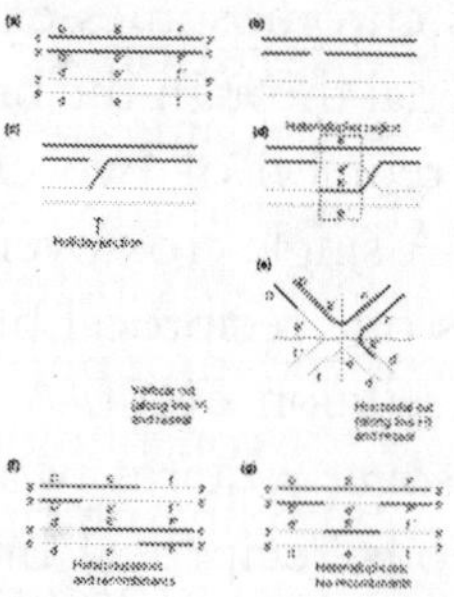

■ crown

the region at the base of the stem of cereals and forage species from which tillers or branches arise. In woody plants, the root-stem junction. In forestry, the top portions of the tree.

■ crown gall

A bulbous growth that occurs at the base of certain plants as result of infection, especially by *Agrobacterium tumefaciens*; a bacterial gene carried by the Ti plasmid is transferred by the bacteria into a higher plant cell, where it causes a tumour-like growth. a.k.a. crown gall tumour. *See Agrobacterium;* hairy root disease.

■ cryobiological preservation

the preservation of germplasm resources in a dormant state by cryogenic techniques, as currently applied to storage of plant seeds and pollen, microorganisms, animal sperm, and tissue culture cell lines. *See ex situ* conservation; gene bank.

■ cryogenic

at very low temperature.

■ cryoprotectant

compound preventing cell damage during freezing and thawing processes. Cryoprotectants are agents with high water solubility and low toxicity. Two types of cryoprotectant agent are commonly used: permeating (glycerol and DMSO (q.v.)) and non-permeating (sugars, dextran, ethylene glycol, polyvinyl pyrolidone and hydroxyethyl starch).

■ cryptic

1. structurally heterozygous individuals not identifiable on the basis of abnormal meiotic-chromosome pairing configurations ('cryptic structural hybrids').
2. a form of polymorphism controlled by recessive genes ('cryptic polymorphism').
3. any mutation which is exposed by a sensitizing mutation and otherwise poorly detected (such mutations probably escape detection because of the plasticity of composition of the corresponding polypeptide).
4. phenotypically very similar species (cryptic species) which do not hybridise under normal conditions.
5. cryptic genetic variation refers to the existence of, for example, alleles conferring high performance for a trait, in a breed that has low performance for that trait.

■ cultivar

a category of plants that are, firstly, below the level of a sub-species taxonomically, and, secondly, found only in cultivation. It is an international term denoting certain cultivated plants that are clearly distinguishable from others by stated characteristics and that retain their distinguishing characters when reproduced under specific conditions.

■ culture

a population of plant or animal cells or micro-organisms that is grown under controlled conditions.

■ culture alteration

a term used to indicate a persistent change in the properties of a culture's behaviour (e.g., altered morphology, chromosome constitution, virus susceptibility, nutritional requirements, proliferative capacity, etc.). The term should always be qualified by a precise description of the change which has occurred in the culture.

■ culture medium

any nutrient system for the cultivation of cells of plants, bacteria or other organisms; usually a complex mixture of organic and inorganic nutrients. *cf* medium.

■ culture room

room for maintaining cultures and often in a controlled environment of light, temperature and humidity. *cf* growth cabinet; incubator.

■ curing

the elimination of a plasmid from its host cell. Many agents which interfere with DNA replication, e.g., ethidium bromide, can cure plasmids from either bacterial or eukaryotic cells.

■ cut

slang: to make a double-stranded break in DNA, usually with a type II restriction endonuclease. E.g., "The DNA was cut with *Eco*RI and run out on a 1% agarose gel." *cf* nick; cleave.

■ cuticle

layer of cutin or wax on the outer surface of leaves and fruits and that reduces water loss.

FORMATION OF NEW EXOSKELETON

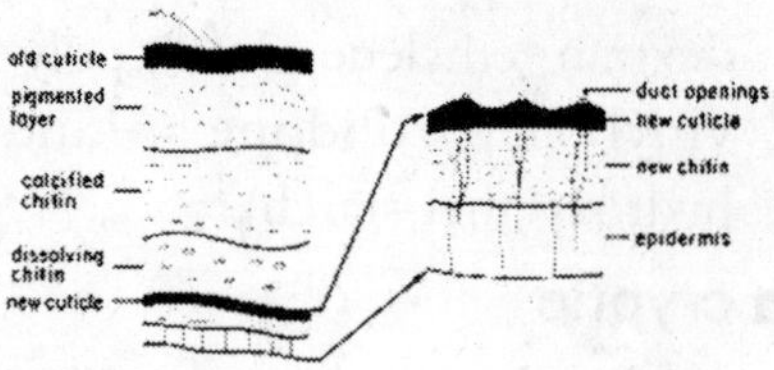

■ cutting

noun: A detached plant part that under appropriate cultural conditions can regenerate the complete plant without a sexual process.

■ cybrid

a cytoplasmic hybrid, originating from the fusion of a cytoplast (the cytoplasm without nucleus) with a whole cell, as in nuclear transfer (although the term is not used in that context). Note that the nucleus and cytoplasm of the fused cell products are from different genetic sources.

■ cyclic AMP

a 'messenger' that regulates many intracellular reactions by transducing signals from extracellular growth factors to cellular metabolic pathways.

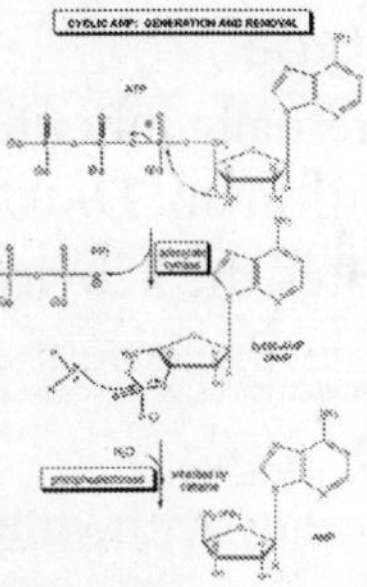

■ **cyclodextrin**
cyclic polymer of dextrose.

■ **cystein**
an amino acid.

■ **cytogenetics**
area of biology concerned with chromosomes and their implications in genetics, cellular activity and variability.

■ **cytokine**
in immunology, any of many soluble molecules that cells produce to control reactions between other cells.

■ **cytokinesis**
cytoplasmic division and other changes exclusive of nuclear division that are a part of mitosis or meiosis.

■ **cytokinin**
plant growth regulators (hormones) characterised as substances that induce cell division and cell differentiation (e.g., BA, kinetin, and 2-iP). In tissue culture, these substances are associated with enhanced callus and shoot development. The compounds are derivatives of adenine.

■ **cytology**
the study of the structure and function of cells.

■ **cytolysis**
cell disintegration.

■ **cytoplasm**
the living material of the cell, exclusive of the nucleus, consisting of a complex protein matrix or gel. The part of the cell in which essential membranes and cellular organelles (mitochondria, plastids, etc.) reside.

■ **cytoplasmic genes**
DNA-containing bodies in the cell but external to the nucleus.

■ **cytoplasmic inheritance**
hereditary transmission dependent on the cytoplasm or structures in the cytoplasm rather than the nuclear genes; extrachromosomal inheritance. Thus, plastid characteristics in plants are inherited by

a mechanism independent of nuclear genes.

■ **cytoplasmic male sterility**

genetic defect due to defective functions of mitochondria in the pollen. Fertilisation will not occur. Exploited in certain plant breeding strategies, such as F_1-hybrid maize cultivars.

■ **cytoplasmic organelles**

discrete sub-cellular structures located in the cytoplasm of cells; these allow division of labour within the cell.

■ **cytosine**

a pyrimidine base found in DNA and RNA.

■ **cytosol**

The fluid portion of the cytoplasm, i.e., the cytoplasm minus its organelles.

Cytosol with Free Ribosomes

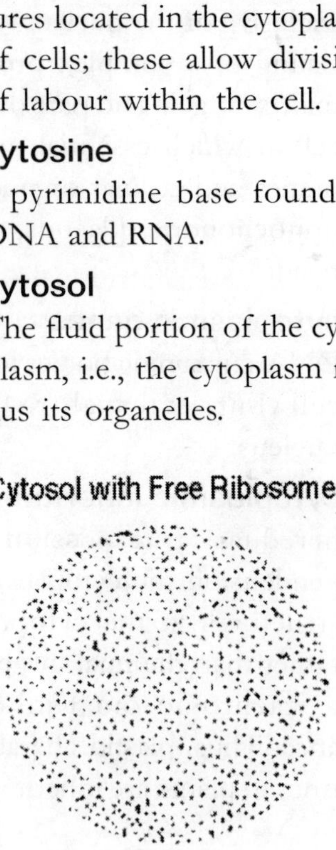

■ **cytotype**

a maternally inherited cellular condition in *Drosophila* that regulates the activity of transposable P elements.

■ **dabs**

antibodies with only one (instead of two) protein chain derived from only one of the domains of the antibody structure. Dabs exploit the finding that, for some antibodies, half of the antibody molecule binds to its target antigen almost as well as the whole molecule. The potential advantages of dabs are that they can be made easily by bacteria or yeasts, and offer a way to clone antibody-like molecules into bacteria, and hence to be able to easily screen millions of antibodies. Related ideas are single-chain antigen binding technology (SCA), biosynthetic antibody binding sites (BABS), minimum recognition units (MRUs), and complementary determining regions (CDRs).

■ **dalton**

a unit of atomic mass roughly equivalent to the mass of a

hydrogen atom. 1.67×10^{-24} g. Named after the famous nineteenth-century chemist, John Dalton (1766-1844)). Used in shorthand expressions of molecular weight, especially as kilo- (kDa) or megadaltons (MDa), which are equal to respectively to 1×10^3 and 1×10^6 daltons.

■ Darwinian cloning
selection of a clone from a large number of essentially random starting points, rather than isolating a natural gene or making a carefully designed artificial one. Molecules which are more similar to those needed are selected, mutated to generate new variants, and re-selected. The cycle proceeds until the required molecule is found. The advantage of the system is that the selection is from a vast number of possibilities.

■ de novo
arising, anew, afresh, once more. Also *ex novo*.

■ death phase
the final growth phase, during which nutrients have been depleted and cell number decreases.

■ deceleration phase
the phase of declining growth rate, following the linear phase and preceding the stationary phase in most batch-suspension cultures.

■ de-differentiation
the process by which cells lose their specialization and proliferate by cell division to form a mass of cells which, in response to appropriate stimuli, may later differentiate again to form either the same cell type or a different one. De-differentiation occurs in response to wounding and in tissue cultures..

■ deficiency
insufficiency or absence of one or more usable forms of enzymatic, nutritional or environmental requirements, so that development, growth or physiological functions are affected.

■ defined
1.fixed conditions of medium, environment and protocol for growth.
2. precisely known and stated elements of a tissue culture medium. *cf* undefined.

■ **degeneracy**

the specification of one amino acid by more than one codon. It arises from the inevitable redundancy resulting from 64 triplets in a triplet code (4 x 4 x 4 = 64) encoding only 20 amino acids.

■ **degeneration**

1.changes in cells, tissues or organs due to disease.

2. the reduction in size or complete loss of organs during evolution.

■ **dehalogenation**

the removal of halogen atoms (chlorine, iodine, bromine, fluorine) from molecules, usually during biodegradation.

■ **dehiscence**

the spontaneous and often violent opening of a fruit, seed pod or anther to release and disperse the seeds or pollen.

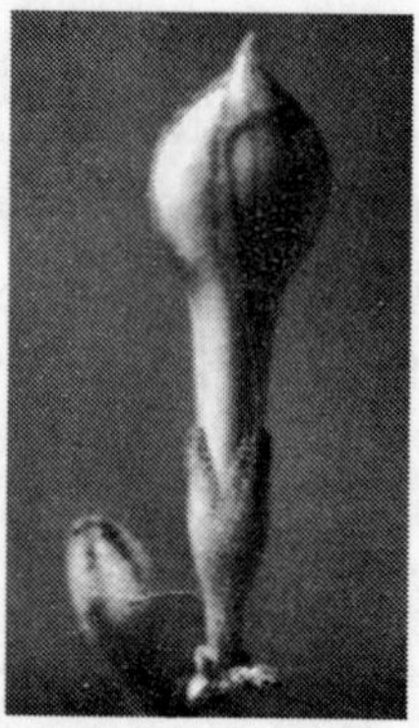

■ **dehydrogenase**

an enzyme that catalyses the remove of hydrogen atoms in biological reactions.

■ **dehydrogenation**

a chemical reaction in which hydrogen is removed from a compound.

■ **de-ionised water**

water which is free of most inorganic (not completely free, since Na is present in ample quantities) and most organic compounds.

■ **deletion**

a mutation involving the removal of one or more base pairs in DNA sequence. Large deletions are visible as the lack of chromosomal segments.

■ **deliberate release**

putting something into the outside world; in biotechnology it means putting a genetically modified organism (GMO) into field trials.

■ **deme**

a group of organisms in the same taxon.

■ **de-mineralise**

to remove the mineral content (salts, ions) from a substance, especially water. Removal methods include distillation and electrodialysis. The process is de-mineralisation.

■ denaturated DNA

duplex DNA that has been converted to single strands by breaking the hydrogen bonds of complementary nucleotide pairs. Usually achieved by heating.

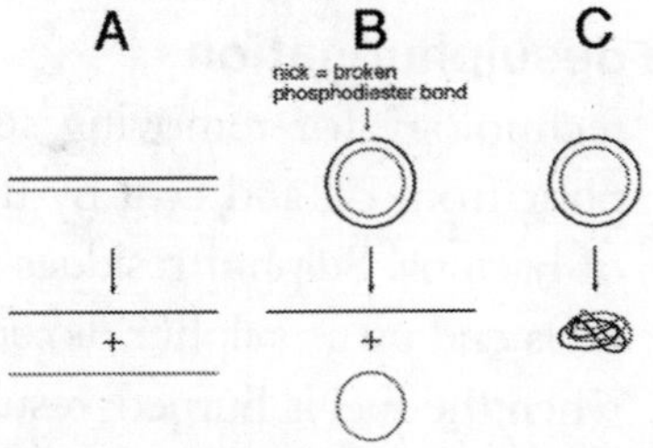

■ denaturation

loss of native configuration of a macro-molecule (protein or nucleic acid) by physical or chemical means, usually accompanied by loss of biological activity. Denatured proteins often unfold their polypeptide chains and express changed properties of solubility. The separation of duplex nucleic acid molecules into single strands. Most commonly used by genetic engineers to describe the destruction of hydrogen bonds maintaining the double-stranded nature of all or part of a DNA molecule.

■ denature

to induce structural alterations that disrupt the biological activity of a molecule. Often refers to breaking hydrogen bonds between base pairs in double-stranded nucleic acid molecules to produce single-stranded polynucleotides, or altering the secondary and tertiary structure of a protein, destroying its activity.

■ denitrification

a chemical process in which nitrates in the soil are reduced to molecular nitrogen, which is released to the atmosphere.

■ density gradient centrifugation

high-speed centrifugation in which molecules "float" at a point where their density equals that in a gradient of caesium chloride or sucrose. The density gradient may either be formed before centrifugation by mixing two solutions of different density (as in sucrose density gradients) or it can be formed by the process of centrifugation itself (as in CsCl and Cs_2SO_4 density gradients).

■ **deoxyribonuclease**
any enzyme that hydrolyses DNA.

■ **de-repression**
the process of "turning on" the expression of a gene or set of genes whose expression has been repressed (turned off). Displacement of a repressor protein from a promoter region of DNA. When attached to the DNA, the repressor protein prevents RNA polymerase from initiating transcription. The "turning on" of a gene.

■ **derivative**
1.resulting from or derived from.
2. term used to identify a variant during meristematic cell division.

■ **desiccant**
any compound used to remove moisture or water.

■ **desiccate**
to dry, exhaust or deprive of water or moisture. Any chemical used for this purpose is called a desiccant. An apparatus for drying and preventing hygroscopic samples from rehydrating is a desiccator. The process is desiccation.

■ **desoxyribonucleic acid**
obsolete spelling of deoxyribonucleic acid (DNA).

■ **dessicator**
apparatus for drying or depriving of moisture.

■ **desulphurisation**
technology for removing sulphur from oil and coal by use of bacteria. Sulphur residues in fuels end up as sulphur dioxide when the fuel is burned, resulting in acid rain. Bacteria may oxidise sulphites (insoluble) into sulphates (soluble), which can be washed away with the bacteria.

■ **detergent**
substance which lowers the surface tension of a solution, improving its cleaning properties (e.g., Tween-20TM, a surfactant and wetting agent). *See* surfactant; wetting agent.

■ **determinate growth**
growth determined and limited in time, as in most floral meristems and leaves. The dif-

ferentiation process is irreversibly established. Determinate growth contrasts with the usual culture growth, which is infinite and indeterminate.

■ determination

process by which undifferentiated cells in an embryo become committed to develop into specific cell types, such as neurons, fibroblasts or muscle cells.

■ determined

describing embryonic tissue at a stage when it can develop only as a certain kind of tissue.

■ development

the sum total of events that contribute to the progressive elaboration of an organism. The two major aspects of development are growth and differentiation.

■ deviation

1.in statistics: the difference between an actual observation and the mean of all observations.

2. an alteration from the typical form, function or behaviour. Mutation or stress are the common reasons behind deviation.

■ dextrin

an intermediate polysaccharide compound resulting from the hydrolysis of starch to maltose by amylase enzymes.

■ diabetes

a disease associated with the absence or reduced levels of insulin, which is a hormone essential for the transport of glucose to cells.

■ diagnostic procedure

a test or assay used to determine the presence of an organism, substance or nucleic acid sequence alteration.

■ diakinesis

a stage of meiosis just before metaphase I, in which the separation of homologous chromosomes is almost completed.

■ diazotroph

an organism that can fix atmospheric nitrogen.

■ dicentric chromosome

a chromosome having two centromeres.

■ dichogamy

the condition in which the male and the female reproductive organs of a flower mature at different times, thereby making self-fertilisation improbable or impossible.

■ dicot

see dicotyledon.

■ dicotyledon

a plant with two cotyledons, or seed leaves. One of the two classes of plants in the Angiosperms (the other class is the monocotyledons). Colloquially called a dicot. Examples include many crop plants (potato, pea, beans), ornamentals (rose, ivy) and timber trees (oak, beech, lime).

■ di-deoxynucleotide

a deoxynucleotide that lacks a 3´- hydroxyl group, and is thus unable to form the 3´-5´ phospho-diester bond necessary for chain elongation. Dideoxynucleotides are used in DNA sequencing and the treatment of viral diseases. Also sometimes referred to as didN.

■ differential centrifugation

a method of separating subcellular particles according to their sedimentation coefficients, which are roughly proportional to their size. Cell extracts are subjected to a succession of centrifuge runs at progressively faster rotation speeds. Large particles, such as nuclei or mitochondria, will be precipitated at relatively slow speeds; higher G forces will be required to sediment small particles, such as ribosomes.

■ differentially permeable

referring to a membrane, through which different substances diffuse at different rates; some substances may be unable to diffuse through such a membrane.

■ differentiation

a process in which unspecialised cells develop structures and functions characteristic of a particular type of cell. Development from one cell to many cells, accompanied by a modification of the new cells for the performance of particular functions. In tissue culture, the term is used to describe the formation of different cell types.

■ **diffusion**

the movement of molecules from a region of higher concentration to a region of lower concentration.

■ **digest**

to cut DNA molecules with one or more restriction endonucleases. *See* cleave.

■ **dihaploid**

an individual which arises from a doubled haploid.

■ **dihybrid**

an individual that is heterozygous for two pairs of alleles; the progeny of a cross between homozygous parents differing at two loci.

■ **dimer**

association of two molecules.

■ **dimethyl sulphoxide**

a highly hygroscopic liquid and powerful solvent with little odour or colour. It is an organic co-solvent used in small quantities to dissolve neutral organic substances in tissue culture media preparation. DMSO also has uses as a cryoprotectant.

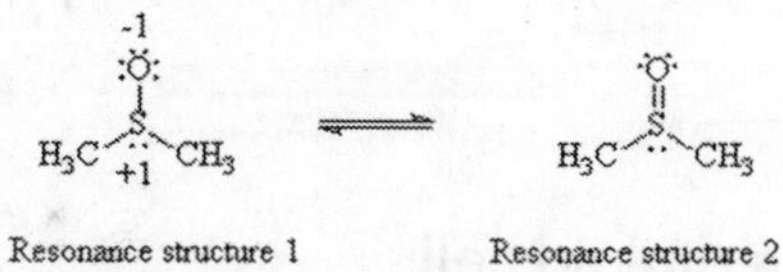

■ **dimorphism**

the existence of two distinctly different types of individuals within a species. An obvious example is the sexual dimorphism in certain animals.

■ **diploid**

1. the status of having two complete sets of chromosomes, most commonly one set of paternal origin and the other of maternal origin.

2. an organism or cell with a double set (2n) of chromosomes (most commonly one of paternal origin, and the other of maternal origin), or referring to an individual containing a double set of chromosomes per cell. Somatic tissues of higher plants and animals are ordinarily diploid in chromosome constitution, in contrast with the haploid gametes.

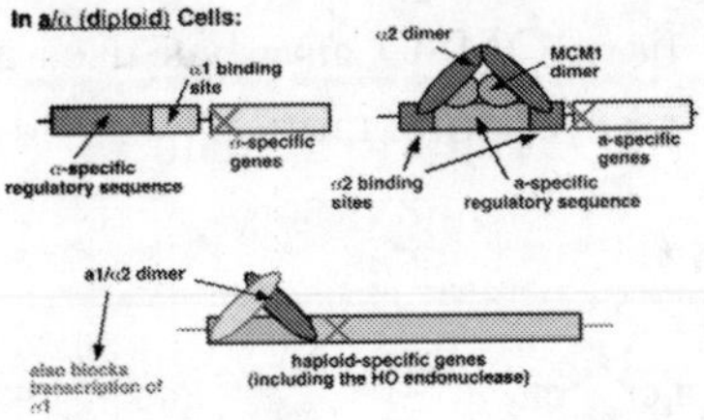

■ **diploid cell**

a cell which contains two sets of chromosomes. *cf* haploid cell.

■ **diplonema**

stage in prophase of meiosis I following the pachytene stage, but preceding diakinesis, in which one pair of sister chromatids begin to separate from the other pair, i.e., the centromeres begin to disjoin.

■ **diplophase**

phase with 2n chromosomes.

■ **direct embryogenesis**

embryoids form directly in culture, without an intervening callus phase, on the surface of zygotic or somatic embryos or on explant tissues (leaf section, root tip, etc.).

■ **direct organogenesis**

formation of organs directly on the surface of cultured intact explants. The process does not involve callus formation.

■ **direct repeat**

two or more stretches of DNA within a single molecule which have the same nucleotide sequence in the same orientation. Direct repeats may be either adjacent to one another or far apart on the same molecule. For example TATTA...TATTA.
ATAAT...ATAAT.

■ **directed mutagenesis**

the process of generation of nucleotide changes in cloned genes by any one of several procedures, including site-specific and random mutagenesis. Also called *in vitro* mutagenesis.

■ **directional cloning**

the technique by which DNA insert and vector molecules are digested with two different restriction enzymes to create non-complementary sticky ends at either end of each restriction fragment, so allowing the insert to be ligated to the vector in a specific orientation and preventing the vector from re-circularising.

■ **disaccharide**

a carbohydrate consisting of two linked sugar units.

Structure of a Disaccharide

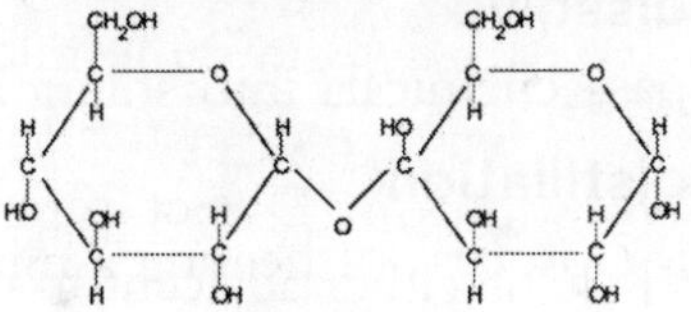

Maltose
(A compound of 2 glucose molecules)

■ **disarm**

to delete from a plasmid or virus genes that are cytotoxic or tumour inducing.

■ **discontinuous variation**

phenotypic variation involving distinct classes, such as red versus white, tall versus dwarf.

■ **discordant**

members of a pair showing different, rather than similar, characteristics.

■ **disease**

the opposite of ease. Any alteration from the state of metabolism necessary for the normal development and functioning of an organism, usually associated with infection by a pathogen or the malfunction or absence of one or more genes.

■ **disease resistance**

the ability to remain healthy by resisting disease or the disease agent. Disease resistance or tolerance is a subject of intense interest in biotechnology.

■ **disease-free**

a plant or animal certified through specific tests as being free of specified pathogens. Disease-free should be interpreted to mean 'free from any known diseases' as 'new' diseases may yet be discovered to be present. *cf* disease indexing.

■ **disease-indexing**

disease-indexed organisms have been assayed for the presence of known diseases according to standard testing procedures. *cf* disease-free.

■ **disinfectation**

full elimination of internal micro-organisms from a culture; disinfectation is rarely obtained. *cf* disinfestation.

■ **disinfestation**

the elimination or inhibition of the activity of surface-adhering micro-organisms. *cf* disinfection.

■ **disjunction**
separation of homologous chromosomes during anaphase I of meiosis; separation of sister chromatids during anaphase of mitosis and anaphase II of meiosis. As soon as the sister chromatids have separated, they are each called a chromosome.

■ **disomy**
the presence of a pair of specific chromosome. This is the normal condition, and abnormal occurrences are monosomy (q.v.), trisomy (q.v.) and nullisomy (with respectively one chromosome of a pair, three or none). There are also abnormal disomic conditions, such as when both chromosomes of the pair were inherited from the same parent.

■ **dispense**
portion out a nutrient medium into containers, such as test tubes, jars, Erlenmeyer flasks, Petri dishes, etc.

■ **dissecting microscope**
a microscope with a low magnifying power of about 50×, used to examine or excise small plant or animal parts.

■ **dissection**
separation of a tissue by cutting for analysis or observation.

■ **dissolve**
pass chemicals into solution.

■ **distillation**
the process of heating a mixture to separate the more volatile from the less volatile parts, and then cooling and condensing the resulting vapour so as to produce a more nearly pure or refined substance.

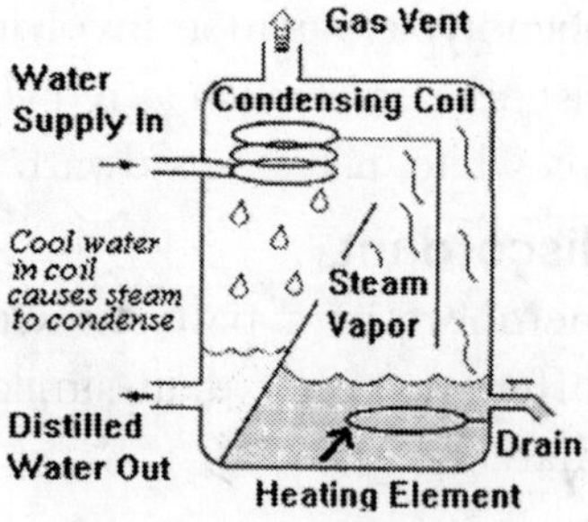

■ **di-sulphide bond**
a chemical bond that stabilises the three-dimensional structure of proteins, and hence the protein's normal function. They form between cysteine residues in the same or different peptide molecules. a.k.a. di-sulphide bridge.

■ **ditype**

in fungi, a tetrad that contains two kinds of meiotic products (spores), e.g., 2AB and 2ab.

■ **diurnal**

term describing the occurrence of an event at least once every 24 hours. *cf* circadian.

■ **dizygotic twins**

two-egg twins, i.e., a pair of individuals that shared the same uterus at the same time, but which arose from separate and independent fertilisation of two ova.

■ **DNA**

deoxyribonucleic acid; formerly spelt desoxyribonucleic acid.The long chain of molecules in most cells that carries the genetic message and controls all cellular functions in most forms of life. The information-carrying genetic material that comprises the genes. DNA is a macro-molecule composed of a long chain of deoxyribonucleotides joined by phospho-diester linkages. Each deoxyribonucleotide contains a phosphate group, the five-carbon sugar 2-deoxribose, and a nitrogen-containing base. The genetic material of most organisms and organelles so far examined is double-stranded DNA; a number of viral genomes consist of single-stranded DNA or single-or double-stranded RNA. In double-stranded DNA, the two strands run in opposite (anti-parallel) directions and are coiled round one another in a double helix. Purine bases on one strand specifically hydrogen bond with pyrimidine bases on the other strand, according to the Watson-Crick rules (A pairs with T; G pairs with C). Hence a constant width for the double helix of 20 Å (2.0 nm) is maintained. In the B-form, DNA adopts a right-handed helical conformation, with each chain making a complete turn every 34 Å (3.4 nm), or once every ten bases.

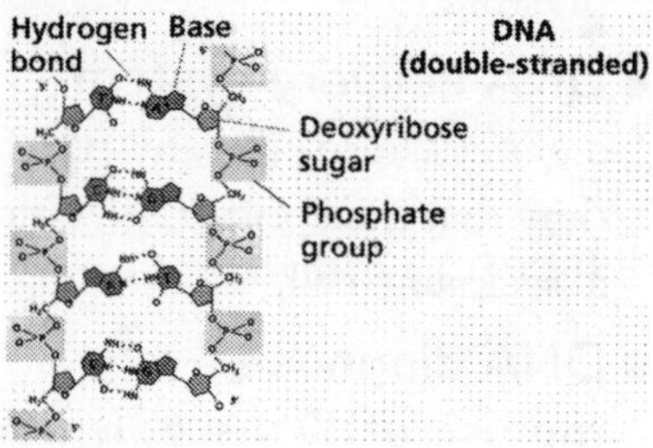

DNA amplification
multiplication of a piece of DNA in a test-tube into many thousands of millions of copies. The most commonly used process is the polymerase chain reaction (PCR) system, but other systems are being developed, including ligase chain reaction (LCR), nucleic acids sequence-dependent amplification, and the Q-b system.

DNA bank
In AnGR: storage of DNA, which may or may not be the complete genome, but should always be accompanied by inventory information. (*Note:* at the present time, animals cannot be re-established from DNA alone.) (Source: FAO, 1999).

DNA construct
a DNA molecule inserted into a cloning vector, usually a plasmid.

DNA delivery system
a generic term for any procedure that transports DNA into a recipient cell.

DNA diagnosis
the use of DNA polymorphisms to detect the presence of a specific allele (often associated with a disease or syndrome) or DNA sequence.

DNA fingerprint
the unique pattern of DNA fragments identified originally by Southern hybridisation (using a probe that binds to a polymorphic region of DNA) or now by polymerase chain reaction (PCR) (using primers flanking the polymorphic region). *See* genetic fingerprinting.

DNA helicase
an enzyme that catalyses the unwinding of the complementary strands of a DNA double helix.

DNA hybridisation
the pairing of two DNA molecules, often from different sources, by hydrogen bonding between complementary nucleotides. This technique is frequently used to detect the presence of a specific nucleotide sequence in a DNA sample.

DNA ligase
an enzyme that catalyses a reaction that links two DNA molecules via the formation

of a phospho-diester bond between the 3´ hydroxyl and 5´ phosphate of adjacent nucleotides. It plays an important role in DNA repair and replication. DNA ligase is one of the essential tools of recombinant DNA technology, enabling (among other things) the incorporation of foreign DNA into vectors. The ligase enzyme encoded by phage T4 is commonly used in gene-cloning experiments. It requires ATP as a co-factor. T4 is used *in vitro* to join the vector and insert DNAs.

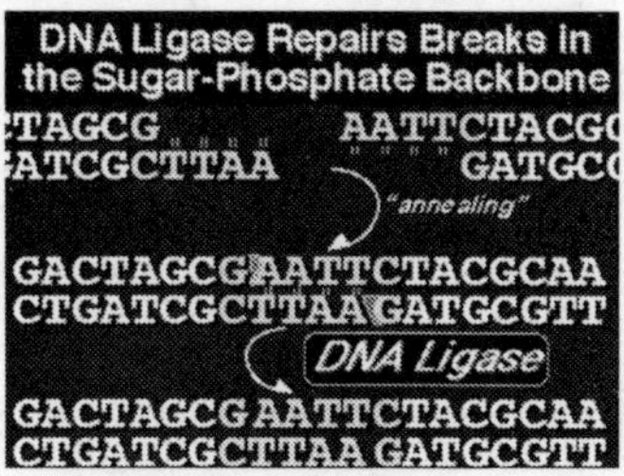

■ DNA micro-array

a small glass surface to which has been fixed an array of DNA fragments, each with a defined location. A typical DNA chip would contain 10 000 discrete spots (each containing a different DNA fragment) in an area of just a few square centimetres. When a solution of fluorescently labelled DNA fragments is hybridised to the chip, spots to which hybridisation occurs are visible as fluorescence. If the spots on the chip are genes (expressed sequence tags, q.v.), hybridisation with cDNA from a particular tissue shows which genes are expressed in that tissue. If the spots are short, synthesised oligonucleotides (approximately 25 bases) corresponding to that part of a gene containing a single nucleotide polymorphisms (SNP)(q.v.), with a separate spot for each of the 4 possible bases at that site, hybridisation with genomic DNA from an individual plant or animal enables that individual to be genotyped at as many SNP loci as are represented on the chip. The big advantage of DNA chips is the extent to which the process of genotyping can be automated, thereby enabling huge numbers of plants or animals to be genotyped for a huge number of loci.

DNA polymerase

an enzyme that catalyses the synthesis of double-stranded DNA, using single-stranded DNA as a template..

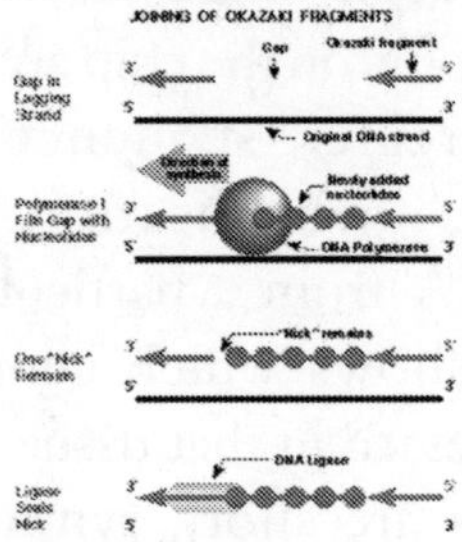

DNA polymorphism

the existence of two or more alternative forms (alleles) of a chromosomal locus that differ in nucleotide sequence or have variable numbers of repeated nucleotide units.

DNA primase

an enzyme that catalyses the synthesis of short strands of RNA that initiate the synthesis of DNA strands.

DNA probe

a labelled (tagged) segment of DNA that is able, after a DNA hybridisation reaction, to detect a specific DNA sequence in a mixture of sequences. If the tagged sequence is complementary to any one in the mixture, the two sequences will form a double helix. This will be identified thanks to its label (either by radioactivity or fluorescence).

DNA repair

a variety of mechanisms that repair errors that occur during DNA replication.

DNA repair enzymes

enzymes that catalyse the repair of DNA.

DNA replication

the process whereby DNA makes exact copies itself, under the action of and control of DNA polymerase.

DNA sequencing

procedures for determining the nucleotide sequence of a DNA fragment. There are two common methods for doing this:

1. the Maxam and Gilbert technique (chemical degradation), that uses different chemicals to break the DNA into fragments at specific bases; or

2. the Sanger technique (called the di-deoxy or chain-terminating method) uses DNA polymerase to make new DNA chains, with di-

deoxy nucleotides (chain terminators) to stop the chain randomly as it grows.

In both cases, the DNA fragments are separated according to length by polyacrylamide gel electrophoresis, enabling the sequence to be read directly from the gel.

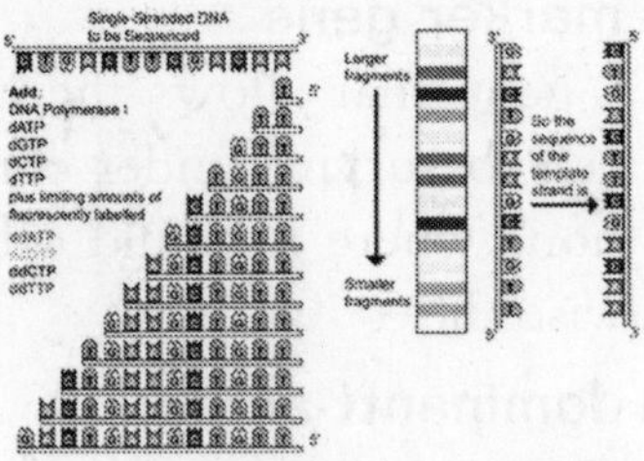

■ DNA topo-isomerase

an enzyme that catalyses the introduction or removal of supercoils in DNA.

■ DNase

an enzyme that catalyses the cleavage of DNA. DNase I is a digestive enzyme secreted by the pancreas, that degrades DNA into shorter nucleotide fragments. Many other endonucleases and exonucleases are involved in DNA repair and replication. *cf* nuclease.

■ Dolly

the [name of the] first mammal to be created by cloning a cell from an adult animal. In this particular case, the cell came from the mammary tissue of an adult ewe. The creation of Dolly showed that the process of differentiation into adult tissue is not, as previously thought, irreversible. The result was achieved by nuclear transfer (q.v.). Dolly's birth was announced in 1997. Since then, cattle and mice have also been cloned from adult cells.

■ domain

a segment of a protein that has a discrete function or conformation. At the protein level, a domain can be as small as a few amino acid residues or as large as half of the entire protein.

■ domestic animal diversity

in AnGR: The spectrum of genetic differences within each breed, and across all breeds within each domestic animal species, together with the species differences; all of which are available for the sustainable intensification of food and agriculture production.

■ dominance

the type of gene action exhibited by a dominant allele (q.v.).

■ dominant

1.describing an allele whose effect with respect to a particular trait is the same in heterozygotes as in homozygotes. The opposite is *recessive*.

2. describing the most conspicuously abundant and characteristic species of a community.

3. describing an animal that is allowed priority in access to food, mates, etc., by others of its species because of its success in previous aggressive encounters.

■ dominant marker selection

selection of cells via a gene encoding a product that enables only the cells that carry the gene to grow under particular conditions. For example, plant and animal cells that express the introduced Neo^r gene are resistant to the compound G418, while cells that do not carry the Neo^r gene are killed by G418. a.k.a. positive selection.

■ dominant selectable marker gene

a gene that allows the host cell to survive under conditions where it would otherwise die.

■ dominant(-acting) oncogene

a gene that stimulates cell proliferation and contributes to oncogenesis when present in a single copy. *See* oncogene.

■ donor plant

an explant, graft or cutting used as a source of plant material for micro-propagation purposes. a.k.a. ortet. *cf* explant.

■ dormancy

an inactive period in the life of an animal or plant during which growth slows or completely ceases. Physiological changes associated with dormancy help the organism survive adverse environmental conditions. Annual plants survive the winter as dormant seeds, while many perennial plants survive as dormant tubers, rhizomes, or bulbs. Hibernation and aestivation in animals help them survive extremes of cold and heat, respectively. *cf* rest period.

■ **dosage compensation**

a phenomenon whereby inactivation of all but one of the X chromosomes in female mammals results in males and females producing the same quantity of peptide from X-linked genes.

■ **double crossing-over**

two simultaneous reciprocal breakage and reunion events between the same two chromatids.

■ **double fertilisation**

a process, unique to flowering plants, in which two male nuclei, which have travelled down the pollen tube, separately fuse with different female nuclei in the embryo sac. The first male nucleus fuses with the egg cell to form the zygote; the second male nucleus fuses with the two polar nuclei to form a triploid nucleus that develops into the endosperm.

■ **double helix**

describes the coiling of the antiparallel strands of the DNA molecule, resembling a spiral staircase in which the paired bases form the steps and the sugar-phosphate backbones form the rails.

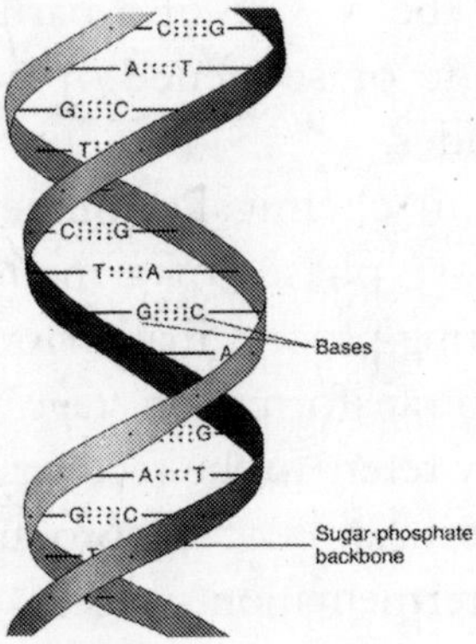

■ **double recessive**

an organism homozygous for a recessive allele at each of two loci.

■ **double-stranded complementary DNA)**

a double-strand DNA molecule created from a cDNA template.

■ downstream

1. in molecular biology, the stretch of nucleotides of DNA that lie in the 3′ direction from the site of initiation of transcription, which is designated as +1 (remembering the convention that the sequence of a DNA molecule is written from the 5′ end to the 3′ end). Downstream nucleotides are marked with plus signs, e.g., +2, +10. Also, to the 3′ side of a particular gene or sequence of nucleotides.

2. in chemical engineering, those phases of a manufacturing process that follow the biotransformation stage. Usually refers to the recovery and purification of the product of a fermentation process.

■ downstream processing

a general term for all the things which happen in a biotechnological process after the biology, be it fermentation of a micro-organism or growth of a plant. It is particularly relevant to fermentation processes, which produce a large quantity of a dilute mixture of substances, products and micro-organisms. These must be separated, and the product must be concentrated and purified, and converted into a form which is useful.

■ drug delivery

method by which a therapeutic agent is delivered to its site of action. For traditional therapeutic agents this is another name for *formulation*. However, biotechnology has allowed the development of a range of new therapeutic-agent delivery systems, such as liposomes and other encapsulation techniques, and a range of mechanisms that target a therapeutic agent to a particular cell or tissue. *See* therapeutic agent.

■ dry weight

the moisture-free weight of tissue obtained by drying at high (oven-drying) or low (freeze-drying) temperatures for an interval sufficient to remove all water.

■ dual culture

a culture made of a plant tissue and one organism (such as a nematode) or an obligate parasite/micro-organism (such as a fungus). Dual culture techniques are used for a variety of purposes, including assessing host-parasite interactions and the production of axenic cultures.

■ duplex DNA
double-stranded DNA.

■ duplication
the occurrence of a segment more than once in the same chromosome or genome.

■ ecdysone
a steroid hormone in insects that stimulates moulting and metamorphosis. It acts on specific genes, stimulating the synthesis of proteins involved in these bodily changes.

■ eclosion
1.emergence of an adult insect from the pupal stage.
2. beginning of germination of fungal spores.

■ ecology
the study of the interactions between organisms and their natural environment, both living and non-living.

■ economic trait locus
a locus influencing a trait that contributes to income. The plural form (economic trait loci) is also abbreviated as ETL.

■ ecosystem
the complex of a living community and its environment, functioning as an ecological unit in nature. *cf* abiotic factors; biotic factors.

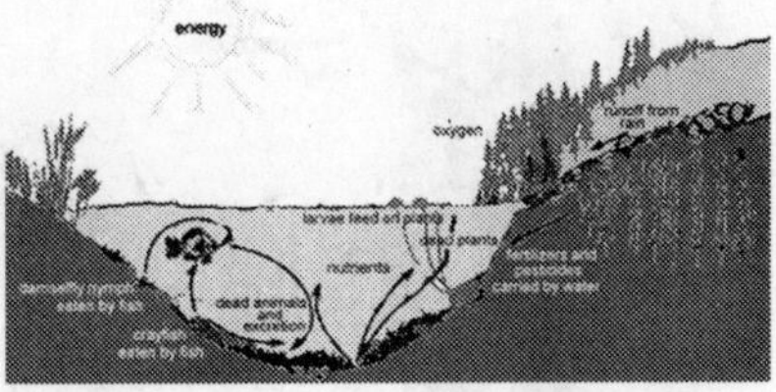

■ ecotype
a population or a strain of an organism that is adapted to a particular habitat.

■ effector cells
cells of the immune system that are responsible for cell-mediated cytotoxicity.

■ effector molecule
a molecule that influences the behaviour of a regulatory molecule, such as a repressor protein, thereby influencing gene expression.

egg

1. the fertilised ovum (zygote) in egg-laying animals after it emerges from the body.
2. the mature female reproductive cell in animals and plants.

elastin

a fibrous protein that is the major constituent of the yellow elastic fibres of connective tissue.

(a) Collagen

electro-blotting

the electrophoretic transfer of macromolecules (DNA, RNA or protein) from a gel, in which they have been separated, to a support matrix, such as a nitrocellulose sheet. A transfer usually used in techniques such as Southern and northern blotting.

electrochemical sensor

type of biosensor in which a biological process is harnessed to an electrical sensor system, such as an enzyme electrode. Other types couple a biological event to an electrical one via a range of mechanisms, such as those based on oxygen and pH. *See* enzyme electrode.

electromagnetic radiation

electromagnetic waves, including ultraviolet (UV), X-rays, and gamma radiation (g rays). Electromagnetic radiation is used to produce mutant cells or organisms, or, in the case of UV, disinfestation and sterilisation, in tissue culture.

electromagnetic spectrum

The range of wavelengths or frequencies over which electromagnetic radiation extends.

electron microscope

a microscope that uses an electron beam focused by magnetic "lenses."

■ **electrophoresis**

a technique that separates charged molecules - such as DNA, RNA or protein - on the basis of relative migration in an appropriate matrix (such as agarose gel or polyacrylamide gel) subjected to an electric field. *See* agarose gel electrophoresis; polyacrylamide gel electrophoresis (PAGE); pulsed-field gel electrophoresis (PFGE).

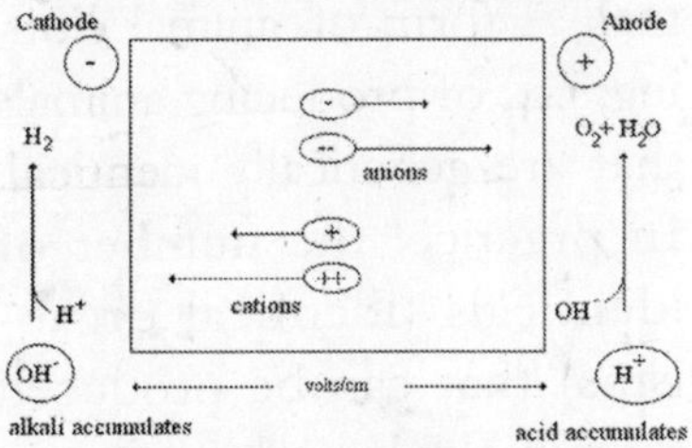

■ **electroporation**

1. an electrical treatment of cells that induces transient pores, through which DNA can enter the cell.

2. the introduction of DNA or RNA into protoplasts or other cells by the momentary disruption of the cell membrane through exposure to an intense electric field.

■ **ELISA**

a sensitive technique for accurately determining specific molecules in a mixed sample. The amount of protein or other antigen in a given sample is determined by means of an enzyme-catalysed colour change, avoiding both the hazards and expense of radioactive techniques.

■ **elongation factors**

soluble proteins that are required for polypeptide chain elongation.

■ **embryo**

an immature organism in the early stages of development. In mammals, this occurs in the first months in the uterus. In plants, it is the structure that develops in the megagametophyte, as result of the fertilisation of an egg cell, or without fertilisation. In aseptic cultures, adventitious embryos show polarization, followed by the growth of a shoot from one end and a root from the other end.

■ **embryo cloning**

the creation of identical copies of an embryo by embryo splitting (q.v.) or by nuclear transfer (q.v.) from undifferentiated embryonic cells.

■ **embryo culture**

the culture of embryos on nutrient media.

■ **embryo multiplication and transfer**

the cloning of animal embryos and their subsequent transfer to recipients (via artificial inembryonation (q.v.)). The cloned embryos can be clones of an embryo or of an adult.

■ **embryo sac**

a large thin-walled space within the ovule of the seed plant in which the egg and, after fertilisation, the embryo develop; the mature female gametophyte in angiosperms. Generally a seven-celled structure. The seven cells are two synergids, one egg cell, three antipodal cells (each with a single haploid nucleus) and one endosperm mother cell with two haploid nuclei.

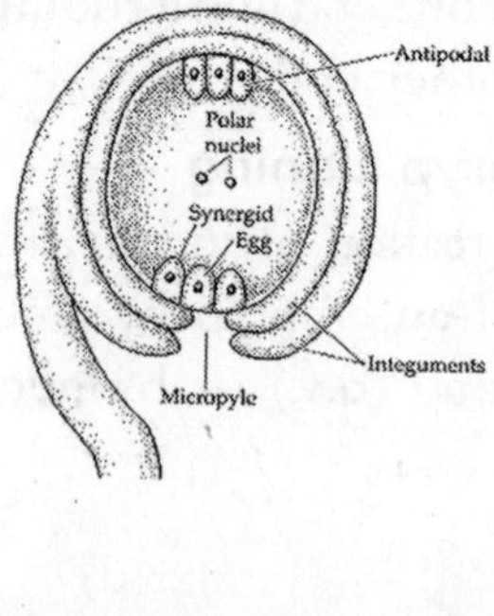

■ **embryo sexing**

the determination of the sex of an embryo, typically by means of PCR (q.v.) involving amplification from a small sample of embryonic tissue, using primers specific for a locus on the Y chromosome.

■ **embryo splitting**

the splitting of young embryos into several sections, each of which develops into an animal. A form of animal cloning, i.e., of producing animals that are genetically identical. In practice, the number of identicals (identical organisms) that can be produced from a single embryo is less than 10.

■ **embryo technology**

generic name for any modification of mammalian embryos. It encompasses embryo cloning, embryo splitting, *in vitro* fertilisation, and embryo storage.

■ **embryogenesis**

1. (general) development of an embryo.

2. (in plants) in vitro formation of plants from plant tissues, through a pathway closely re-

sembling normal embryogeny from the zygote; if this development in culture involves somatic cells and not the zygote, it can be indicated by using the term *adventitious embryogenesis* or *somatic embryogenesis*. The generation of embryos has two stages: initiation and maturation. Initiation needs a high level of the group of plant hormones called auxins; maturation needs a lower level. Other chemicals have to be at suitable levels. The procedure involves the explanting of a piece of plant tissue and putting it on a high-auxin medium, where the cells grow into a mass of callus. This is then transferred to a maturation medium, where the callus starts to initialize organs, ultimately growing a root and a shoot.

■ **embryoid**

an embryo-like body developing *in vitro*. It forms a complete, self-contained platelet with no vascular connection with the callus. The term embryoid is no longer commonly used.

■ **embryonic stem cells**

cells of the early embryo that can give rise to all differentiated cells, including germ line cells.

■ **empirical**

relating to or based upon practical experience, trial and error, direct observation or observation alone, without benefit of scientific method, knowledge or theory.

■ **encapsidation**

the process by which a virus' nucleic acid is enclosed in a capsid. *See* capsid; coat protein.

■ **encapsulating agents**

anything which forms a shell around an enzyme or bacterium, although the agents used are usually polysaccharides such as alginate or agar. The agents are inert and allow nutrients and oxygen to diffuse into and out of the sphere readily, and are easy to convert from gel (solid) to sol (liquid) or solution form by altering the temperature or the concentration of ions.

■ **encapsulation**

any method of getting something, usually an enzyme or

bacterium, into a small package or capsule while it is still working or alive. It is a method for immobilising cells for use in a bioreactor.

■ **encode**

to specify, after decoding by transcription and translation, the sequence of amino acids in a protein.

■ **endangered breed**

In AnGR: a breed where the total number of breeding females is between 100 and 1 000 or the total number of breeding males is less than or equal to 20 and greater than five; or the overall population size is close to, but slightly above 100 and increasing, and the percentage of pure-bred females is above 80%; or the overall population size is close to, but slightly above, 1 000 and decreasing, and the percentage of pure-bred females is below 80%.

■ **endangered species**

a plant or animal species in immediate danger of extinction because its population numbers have reached a critical level or its habitats have been drastically reduced.

■ **endemic**

1. describing a plant or animal species whose distribution is restricted to one or a few localities.
2. describing a disease or a pest that is always present in an area.

■ **end-labelling**

the introduction of a radioactive atom at the end of a DNA or RNA molecule. A commonly used method is to use T4 polynucleotide kinase to introduce a ^{32}P atom onto the end of a DNA molecule.

■ **endocrine gland**

any gland in an animal that manufactures hormones and secretes them directly into the bloodstream to act at distant sites in the body, known as target organs or cells. *cf* exocrine gland.

■ **endocrine interference**

interference with the normal balance hormones.

■ **endocytosis**

the process by which materials enter a cell without passing through the cell membrane. The membrane folds around material outside the cell, resulting in the formation of a saclike vesicle into which the material is incorporated. This vesicle is then pinched off from the cell surface so that it lies within the cell. *See* phagocytosis; pinocytosis.

■ **endoderm**

the internal layer of cells of the gastrula, which will develop into the alimentary canal (gut) and digestive glands of the adult.

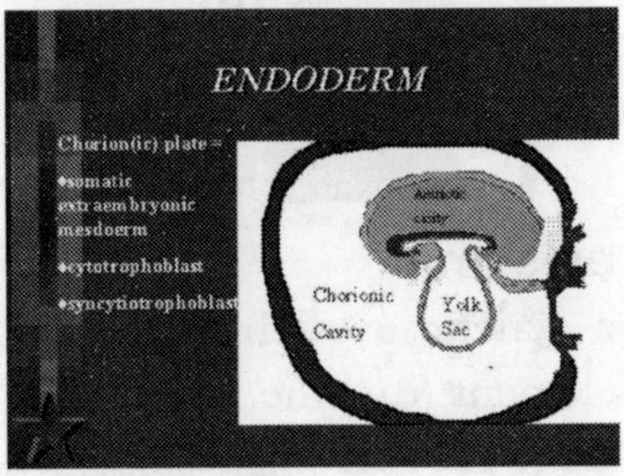

■ **endodermis**

the layer of living cells, with various characteristically thickened walls and no intercellular spaces, which surrounds the vascular tissue of certain plants and occurs in nearly all roots and certain stems and leaves. The endodermis separates the cortical cells from cells of the pericycle.

■ **endogamy**

the fusion of reproductive cells from closely related parents, i.e., inbreeding. *cf* exogamy.

■ **endogenote**

the part of the bacterial chromosome that is homologous to a genome fragment (exogenote) transferred from the donor to the recipient cell in the formation of a merozygote.

■ **endopolyploidy**

the result of nuclear divisions without subsequent cytoplasmic division (cytokinesis); the polyploids so obtained are called endopolyploids. *See* polyploidy.

■ **endoprotease**

an enzyme that cleaves the peptide bonds between amino acids within a protein. Cleavage is usually at one or more specific sites.

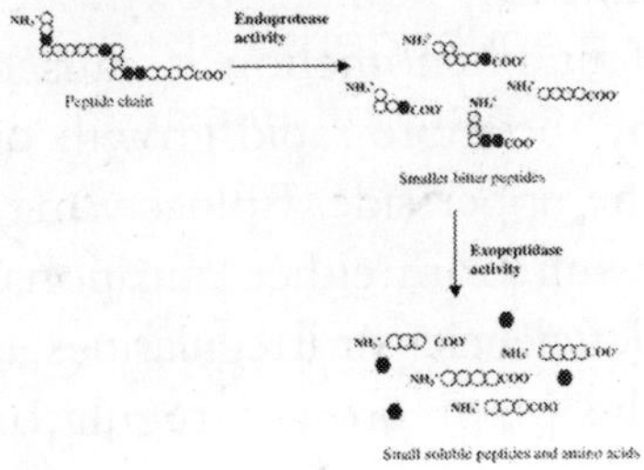

■ **endoreduplication**

chromosome reproduction during interphase. 4-chromatid chromosomes (diplochromosomes) are seen during this phase.

■ **epigenesis**

describes the developmental process whereby each successive stage of normal development is built up on the foundations created by the preceding stages of development; an embryo is built up from a zygote, a seedling from an embryo, and so on.

■ **epigenetic**

a term referring to the non-genetic causes of a phenotype.

■ **epigenetic variation**

non-hereditary and reversible variation; often the result of a change in gene expression.

■ **epinasty**

a process by which the growth of branches or petioles is abnormally pointing downward. This phenomenon is caused by the more rapid growth of the upper side. Epinasty may result from either nutritional deficiencies or irregularities at the plant growth regulator level. Not to be confused with wilting, as epinastic tissues are turgid.

■ **epiphyte**

a plant that grows upon another plant, but is neither parasitic on it nor rooted in the ground.

■ **episome**

a genetic extrachromosomal element (e.g., the fertility factor (F) in *Escherichia coli*) which replicates within a cell independently of the chromosome and is able to integrate into the host chromosome. The step of integration may be governed by a variety of factors and so the term episome has lost favour and been superseded by the wider term *plasmid*. Plasmids and F factors are episomes.

■ **epistasis**

interaction between genes at different loci, e.g., one gene suppresses the effect of another gene that is situated at a different locus. Suppressed genes are said to be hypostatic. Dominance is associated with members of allelic pairs, whereas epistasis is interaction among products of non-alleles.

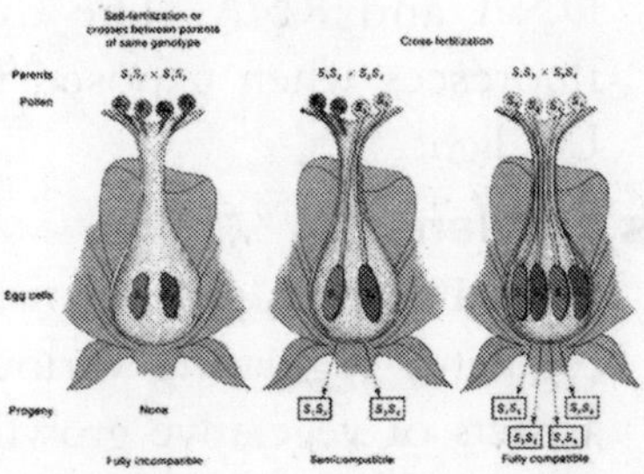

■ **epitope**

a specific chemical domain on an antigen that stimulates the production of, and is recognised by, an antibody. Each epitope on a molecule such as protein elicits the synthesis of a different antibody. a.k.a. antigenic determinant.

■ **epizootic**

a disease affecting a large number of animals simultaneously.

■ **equational division**

mitotic-type division that is usually the second division in the meiotic sequence; somatic mitosis and the non-reductional division of meiosis. A chromosome division in which the two chromatids of each duplicated chromosome separate longitudinally, prior to being incorporated into two daughter nuclei.

■ **equatorial plate**

the figure formed by the chromosomes in the centre (equatorial plane) of the spindle in mitosis.

■ **equilibrium**

a state of dynamic systems in which there is no net change.

■ **equilibrium density gradient centrifugation**

a procedure used to separate macro-molecules based on their density (mass per unit volume).

■ **equimolar**

identical molar concentrations. *See* molarity; mole.

■ **Erlenmeyer flask**

a conical flat-bottomed laboratory flask with a narrow neck, designed by E. Erlenmeyer. Widely used for culturing micro-organisms.

■ escherichia coli

a commensal bacterium inhabiting the colon of many species. *E. coli* is widely used in biology, both as a simple model of cell biochemical function and as a host for molecular cloning experiments. In environmental studies, it is a key indicator of water pollution due to human sewage effluent.

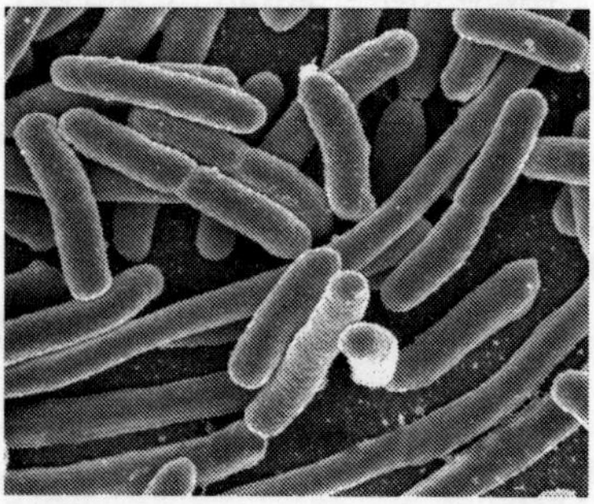

■ ethanol

commonly used to disinfest plant tissues, glassware utensils and working surfaces in tissue culture manipulations. The concentration used is 70% (v/v) for disinfecting and 95% (v/v) when flaming tools. Ethanol is also used to dissolve water-insoluble additions (addendums) to culture media.

■ ethephon

(2-chloroethyl) phosphonic acid ($ClC_2PO_3H_6$; f.w. 144.50) Through a spontaneous degradation of ethephon, ethylene is produced. Ethephon is a synthetic compound commonly used to treat cultured cells or unripened fruit with ethylene (q.v.).

■ ethidium bromide

a fluorescent dye used to stain DNA and RNA. The dye fluoresces when exposed to UV light.

■ ethylene

C_2H_4 a gaseous plant growth regulator regulating various aspects of vegetative growth, fruit ripening and abscission of plant parts.

■ euploid

an organism or cell having a chromosome number that is an exact multiple of the monoploid *(n)* number. Terms used to identify different levels in an euploid series are diploid, triploid, tetraploid, and so on. *cf* aneuploid.

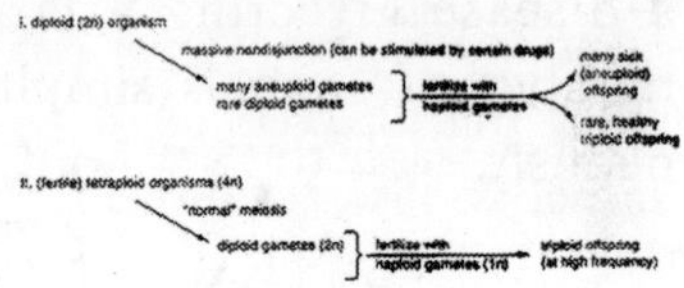

■ **evaluation**

In AnGR: measurement of the characteristics that are important for production and adaptation, either of individual animals or of populations, most commonly in the context of comparative evaluation of the traits of animals or of populations.

■ **evapotranspiration**

the process of water loss in vapour form from a unit surface of land both directly and through leaf surfaces during a specific period of time.

■ **evolution**

the process by which the present diversity of plant and animal life arose from the earliest organisms, a process believed to have been continuing for at least 3 000 million years.

■ **ex vitro**

organisms removed from tissue culture and transplanted; generally plants to soil or potting mixture.

■ **ex vivo gene therapy**

the delivery of a gene or genes to the isolated cells of an individual. After culturing, the transformed cells are introduced back into the individual by transfusion, infusion or injection, to alleviate a genetic disorder.

■ **excinuclease**

the endonuclease-containing protein complex that excises a segment of damaged DNA during excision repair.

■ **excision**

the natural or *in vitro* enzymatic release (removal) of a DNA segment from a chromosome or cloning vector.

■ **excision repair**

DNA repair processes that involve the removal of a damaged or incorrect segment of DNA and its replacement by the synthesis of a new strand using the complementary strand of DNA as template.

■ **extension**

single-stranded DNA region consisting of one or more nucleotides at the end of a strand of duplex DNA. a.k.a. protruding end; sticky end; overhang; cohesive end.

■ **extinct breed**

in AnGR: a breed where it is no longer possible to recreate

the breed population. Extinction is absolute when there are no breeding males (semen), breeding females (oocytes), nor embryos remaining.

■ **extinction**

the irreversible condition of a species or other group of organisms of having no living representatives in the wild, which follows the death of the last surviving individual of that species or group. Extinction may occur on a local or global level; it can result from various human activities, including the destruction of habitats or the overexploitation of species that are hunted or harvested as a resource.

■ **extrachromosomal**

things that are not part of the chromosomes; DNA units in the cytoplasm that control cytoplasmic inheritance.

■ **extrachromosomes**

self-replicative genetic elements separate from main chromosome(s) of a cell. This definition usually excludes viruses, but the division is somewhat arbitrary. In bacteria, plasmids are the principal extrachromosomes; they encode functions which are not essential to the growth and division of the host cell. In eukaryotes, extrachromosomes may be either essential or dispensable. They may inhabit (i) the nucleus, e.g., extrachromosomal rDNA molecules, yeast 2mm plasmid; (ii) the cytosol, e.g., dsRNA molecules in fungi; or (iii) the cytoplasmic organelles, e.g., mitochondrial DNA, chloroplast DNA. Eukaryotic extrachromosomal elements may be recognised genetically by their failure to show segregation at meiosis.

■ **exude**

slowly discharge; leak liquid material (exudate such as tannins or oxidised polyphenols) through pores or cuts, or by diffusion into the medium. In some woody plant species, exudation is associated with a lethal browning of explants.

■ **F factor**

a bacterial episome that confers the ability to function as

a genetic donor in conjugation; the fertility factor in bacteria.

■ F_1

the first filial generation in a cross between any two parents; the first generation of descent from a given mating.

■ F_2

the second filial generation, produced by crossing two members of the F_1, or by self-pollinating the F_1. The "grandchildren" of a given mating.

■ **factorial mating**

a mating scheme in which each male parent is mated with each female parent. Made possible in animals by means of *in vitro* embryo production (q.v.). Such a mating scheme substantially reduces the rate of inbreeding in a selection programme.

■ **FAD**

flavin adenine dinucleotide.A co-enzyme important in various biochemical reactions. It comprises a phosphorylated vitamin B_2 (riboflavin) molecule linked to the nucleotide adenine monophosphate (AMP). It functions as a hydrogen acceptor in dehydrogenation reactions, being reduced to $FADH_2$. This in turn is oxidised to FAD by the electron transport chain, thereby generating ATP (two molecules of ATP per molecule of $FADH_2$).

Flavin adenine dinucleotide
(FAD-Oxidised Form)

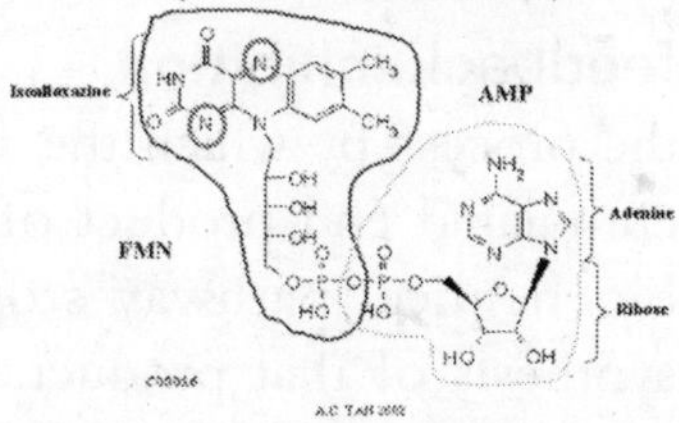

■ **false-negative**

a negative assay result that should have been positive.

■ **false-positive**

a positive assay result that should have been negative.

■ **farm animal genetic resources (AnGR)**

in AnGR: those animal species that are used, or may be used, for the production of food and agriculture, and the populations within each of them. These populations within each species can be classified as wild and feral populations, landraces and primary populations, standardized breeds, selected lines, and any conserved genetic material. (Source: FAO, 1999).

■ fed-batch fermentation

culture of cells or micro-organisms where nutrients are added periodically to the bioreactor.

■ feedback inhibition

the process by which the accumulated end product of a biochemical pathway stops synthesis of that product. A late metabolite of a synthetic pathway regulates synthesis at an earlier step of the pathway. *cf* end-product inhibition.

■ fermentation

the anaerobic breakdown by micro-organisms of complex organic substances, especially carbohydrates like glucose. The process is energy-yielding. Fermentation is often misused to describe large-scale aerobic cell culture in specialised vessels (fermenters, bioreactors) for secondary product synthesis.

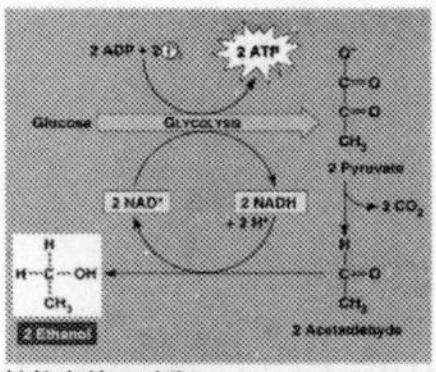

(a) Alcohol fermentation

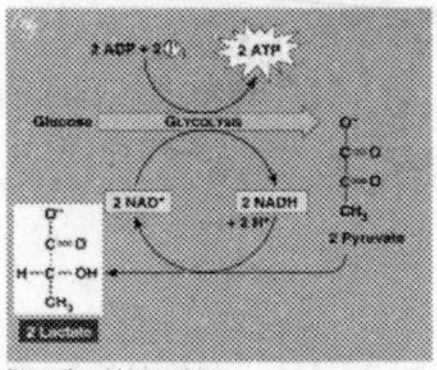

(b) Lactic acid fermentation

■ fermentation substrates

materials used as food for growing micro-organism. The fermentation substrates and the trace materials needed, together with chemicals added to make the fermentation easier, form the culture medium.

■ fertile

of an organism: capable of breeding and reproduction.

■ fouling

the coating or plugging (by materials or micro-organisms) of equipment, thus preventing it from functioning properly.

■ founder animal

in transgenesis research, an organism that carries a transgene in its germ line and can be used in matings to establish a pure-breeding transgenic line, or one that acts as a breeding stock for transgenic animals.

■ **founder principle**

the possibility that a new, small, isolated population may be genetically different from the "parent" population, because the founding individuals (being a small, random sample from the large, "parent" population) could be quite different from typical members of the "parent" population.

■ **four-base cutter**

a type II restriction endonuclease that binds (and subsequently cleaves) DNA at sites that contain a sequence of four nucleotide pairs that is uniquely recognised by that enzyme. Because any sequence of four bases occurs more frequently by chance than any sequence of six bases, four-base cutters cleave more frequently than do six-base cutters. Thus, four-base cutters create smaller fragments than six-base cutters.

■ **fractionation**

separation.

■ **frameshift mutation**

a mutation that changes the reading frame of an mRNA, either by inserting or deleting nucleotides.

■ **free water**

water released by a cell when freezing occurs in intercellular spaces. *cf* bound water.

■ **free-living conditions**

natural or greenhouse conditions where the plantlets are transferred from *in vitro* conditions to soil mixtures. In such instances, plantlets must manufacture their own food supply for survival.

■ **freeze-drying**

the process of drying a tissue or an organ in a frozen state under vacuum. Tissues are freeze-dried to measure their dry weight or to preserve them for future analysis. Freeze-drying is the standard way of preserving micro-organisms for long periods of time. *cf* lyophilisation.

■ **frequency distribution**

a graph showing either the relative or absolute incidence

of classes in a population. The classes may be defined by either a discrete or a continuous variable; in the latter case, each class represents a different interval on the scale of measurement.

■ fresh weight

the weight, including the water content, of a plant or plant part at the time of harvest. *cf* wet weight.

■ friable

a term commonly used to describe a crumbling or fragmenting callus. A friable callus is easily dissected and readily dispersed into single cells or clumps of cells in solution.

■ fungicide

an agent, such as a chemical, that kills fungi.

■ furfuraldehyde

used industrially as a solvent and as a raw material for synthetic resin.

■ fusion gene

a hybrid gene created by joining portions of two different genes (to produce a new protein) or by joining a gene to a different promoter (to alter or regulate gene transcription).

■ gelatin

a glutinous, proteinaceous gelling and solidifying agent. Gelatin is produced by boiling animal connective tissues, which partially hydrolyses the collagen. Gelatin is used to gel or solidify nutrient solutions for tissue culture. *cf* agar.

■ gelatinisation

steam cooking of milled grain, a process that increases the surface area of the starch and converts the original mash to a material with a gel-like consistency.

■ GelriteTM

the brand name of a *Pseudomonas*-derived refined polysaccharide used as a gelling agent and agar substitute.

■ GEM

genetically engineered microorganism.

■ gene addition

the addition of a functional copy of a gene to the genome of an organism.

■ **gene amplification**

the selective production of multiple gene copies without a proportional increase in others.

■ **gene bank**

the physical location where collections of genetic material in the form of seeds, tissues or reproductive cells of plants or animals are stored.

■ **genetic complementation**

when two DNA molecules that are in the same cell together produce a function that neither DNA molecule can supply on its own. a.k.a. complementation.

■ **genetic disease**

a disease that has its origin in changes to the genetic material. Usually refers to diseases that are inherited in a Mendelian fashion, although non-inherited forms may also result from DNA mutation.

■ **genetic distance**

in AnGR: a measure of the genetic similarity between any pair of populations. Such distance may be based on phenotypic traits, allele frequencies or DNA sequences. For example, genetic distance between two populations having the same allele frequencies at a particular locus, and based solely on that locus, is zero. The distance for one locus is maximum when the two populations are fixed for different alleles. When allele frequencies are estimated for many loci, the genetic distance is obtained by averaging over these loci.

■ **genetic distancing**

in AnGR: the collection of the data on phenotypic traits, marker allele frequencies or DNA sequences for two or more populations, and estimation of the genetic distances between each pair of populations. From these distances, the best representation of the relationships among all the populations may be obtained. (Source: FAO, 1999).

■ **genetic diversity**

the heritable variation within and among populations which is created, enhanced or maintained by evolutionary forces.

genetic drift
change in allele frequency from one generation to another within a population, due to the sampling of finite numbers of genes that is inevitable in all real (finite) populations. The smaller the population, the greater is the genetic drift. Sooner or later (depending on the size of the population), genetic drift results in loss of alleles from a population, and hence leads to a loss of genetic variation. Because of this, the minimization of genetic drift is an important consideration for conservation of genetic resources. *See* genetic diversity.

genetic engineering
changes in the genetic constitution of cells (apart from selective breeding) resulting from the introduction or elimination of specific genes through modern molecular biology techniques. This technology is based on the use of a vector for transferring useful genetic information from a donor organism into a cell or organism that does not possess it.

germ layers
the layers of cells in an animal embryo at the gastrula stage, from which the various organs of the animal's body will be derived.

germ line
a lineage of 'generative' cells (= germ track) ancestral to the gametes (sperm and egg cells proper) which, during the development of an organism (animal or plant), are set aside as potential gamete-forming tissues. These ancestral cells, together with the gametes, are called germ cells, as opposed to somatic cells. Location, nature and time of formation of potential gamete-forming tissues are species specific, and may vary greatly from one species to another.

germ line cells
cells that produce gametes.

germ line gene therapy
the delivery of a gene or genes to a fertilised egg or an early embryonic cell. The

transferred gene(s) is present in all or some of the nuclei of the cells of the mature individual, including possibly the reproductive cells, and alters the phenotype of the individual that develops.

■ germplasm
the genetic material that forms the physical basis of hereditary and which is transmitted from one generation to the next by means of the germ cells.

■ graft inoculation test
a test based on the use of a suspected viral carrier which is grafted to an indicator plant. If symptoms appear in the indicator plant, the viral assay is positive.

■ graft union
the point at which a scion from one plant is joined to a stock from another plant.

■ Gram staining
a technique to distinguish between two major bacterial groups, based on stain retention by their cell walls. Bacteria are heat-fixed, then stained with crystal violet, followed by iodine solution, and then rinsed with alcohol or acetone. Gram-positive bacteria are stained bright purple, while Gram-negative bacteria are decolourised.

■ Gram-negative organism
any prokaryotic organism that does not retain the first stain (crystal violet) used in Gram's staining technique. It does retain the second stain (safranin O) and therefore has a pink colour when viewed under a light microscope. Retention of the stain is due to the structure of the cell wall.

■ Gram-positive organism
any prokaryotic organism that retains the first stain used in the Gram technique, which gives a purple-black colour when viewed under a light microscope. Retention of the stain is due to the structure of the cell wall.

■ **haploid cell**

a cell containing only one set, or half the usual (diploid) number, of chromosomes.

■ **haplotype**

1. a group of alleles, each from a different locus in the same region of a chromosome, that exist in the same double helix.

■ **haptoglobin**

a serum protein, alpha globulin, that interacts with haemoglobin during recycling of the iron molecule of haemoglobin.

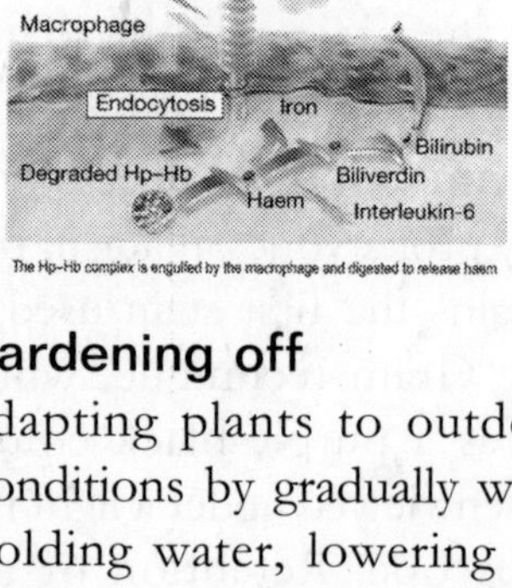

The Hp–Hb complex is engulfed by the macrophage and digested to release haem

■ **hardening off**

adapting plants to outdoor conditions by gradually withholding water, lowering the temperature, increasing light intensity, or reducing the nutrient supply. The hardening-off process conditions plants for survival when transplanted outdoors. The term is also used for gradual acclimatisation to *in vivo* conditions of plants grown *in vitro*, e.g., gradual decrease in humidity. *cf* acclimatisation; free-living conditions.

■ **Hardy-Weinberg equilibrium**

the frequencies of genotypes at a locus resulting from random mating at that locus; for two alleles, A_1 and A_2, with respective frequencies p and q, the Hardy-Weinberg equilibrium frequencies are p^2 A_1A_1, $2pq$ A_1A_2, q^2 A_2A_2. Despite the simplifying assumptions required to predict these frequencies, most loci in most populations are in Hardy-Weinberg equilibrium. Thus the Hardy-Weinberg law, which predicts these frequencies, is one of the great unifying themes of biology.

■ **hemizygous**

the condition in which only one allele of a pair is present, as in sex linkage or as a result of deletion. Genes present only once in the genotype and not in pairs; as in haploids, in differential segments of sex chromosomes, or in diploids as a result of aneuploidy or loss of chromosome segments.

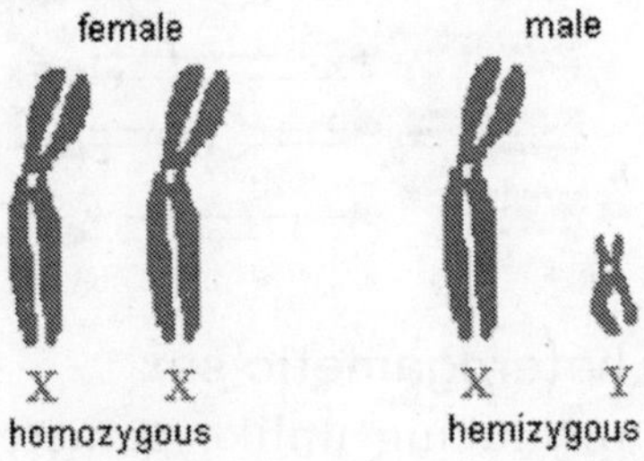

■ HEPA filter

a filter capable of screening out particles larger than 0.3 μm. HEPA filters are used in laminar air flow cabinets (hoods) for sterile transfer work.

■ herbicide

any substance that is toxic to plants; usually applied to agrochemicals intended to kill specific unwanted plants, such as weeds.

■ herbicide resistance

the ability of a plant to withstand herbicide. Herbicide resistance has been one of the early targets of plant genetic engineering. If a herbicide is sprayed onto a field planted with such resistant crops, then all the plants except the crop would be killed, thus providing an effective method of weed control without having to develop herbicides specific to each weed type. There is substantial concern in some quarters about the widespread use of this technology, which is essentially giving the plant kingdom the ability to evade man's most effective herbicides. The concern are that, firstly, such engineering will lead to increased use of the herbicides, at a time when it is generally accepted that the use of chemicals should be kept as low as possible, and that, secondly, there is the possibility that resistant crop plants will escape to become weeds, or that their resistant genes could be transferred to other species, including weeds.

■ heredity

resemblance among individuals related by descent; transmission of traits from parents to offspring.

■ **heterochromatin**

regions of chromosomes that stain darkly even during interphase; thought to be for the most part genetically inactive. *cf* euchromatin.

■ **heteroduplex**

a double-stranded DNA molecule or DNA-RNA hybrid, where each strand is of a different origin, and consequently containing one or more mismatched (non-complementary) base pairs. A DNA duplex is prepared by the hybridisation of single-stranded DNA molecules derived from two different sources. Where the two DNAs have identical or very similar sequences, a double-stranded molecule will be established, whereas where the two DNAs differ in sequence, single-stranded regions will remain. A heteroduplex will be revealed as single-strand "bushes" when DNA is observed electron microscopically. A map of homologous and non-homologous regions of the two molecules may thereby be constructed. This process is known as heteroduplex mapping.

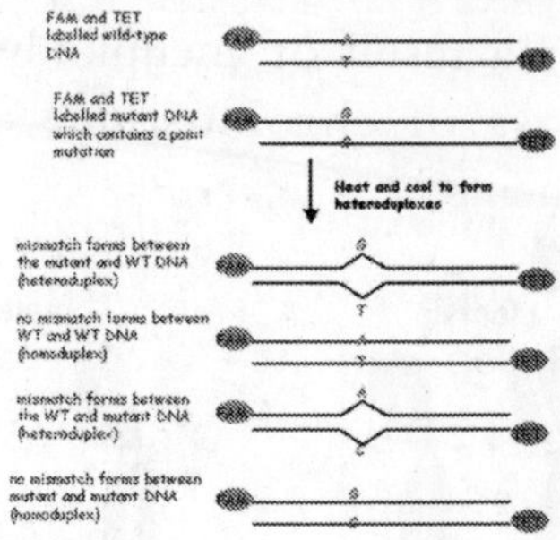

■ **heterogametic sex**

producing unlike gametes with regard to the sex chromosomes. In mammals, the XY male is heterogametic, and the XX female is homogametic.

■ **heterogeneous nuclear RNA**

large RNA molecules, which are unedited mRNA transcripts, or pre-mRNAs found in the nucleous of a eukaryotic cell.

■ **heterokaryon**

cell with two or more different nuclei as a result of cell fusion. Opposite: homokaryon.

■ **heterologous**

from a different source, as in heterologous DNA.

■ heterologous probe
a DNA probe that is derived from one species and used to screen for a similar DNA sequence from another species.

■ heteroplasmy
a cellular condition in which two genetically different types of organelles are present. *cf* homoplasmy.

■ heteroploid
term given to a cell culture when the cells comprising the culture possess nuclei containing chromosome numbers other than the diploid number.

■ heteropyknosis
property of certain chromosomes, or of their parts, to remain more dense during the cell cycle and to stain more intensely than other chromosomes or parts.

■ heterozygote
(Gr. *heteros*, different + *zygon*, yoke) an individual that has different alleles at the same locus in its two homologous chromosomes.

■ Hfr
high-frequency recombination strain of *Escherichia coli;* in such strains, the F episome is integrated into the bacterial chromosome.

■ histocompatibility
the degree to which tissue from one organism will be tolerated by the immune system of another organism.

■ histocompatibility complex
the collection of genes coding for peptides present on the surface of nucleated cells; these peptides are responsible for the differences between genetically non-identical individuals that cause rejection of tissue grafts between such individuals. These peptides were originally called histocompatibility antigens. They are now called histoglobulins, reflecting their structural similarity to immunoglobulins.

■ hollow fibre
a tube of a porous material, having an internal diameter of a fraction of a millimetre, and so its ratio of surface area to internal volume is very large. This has had two types of application. Firstly, hollow fibres can be used as filters.

Because they have a huge surface area, they take much longer to clog up than normal filters. Secondly, they are used in the hollow fibre bioreactor, in which cells are kept inside the hollow, porous fibres, and the culture medium is circulated outside the reactor. The fibres let nutrients in and products out (as they are in solution), but do not allow the passage of cells. Hollow fibre bioreactors are very effective for maintaining mammalian cells in culture because they have a very large surface area for the cells to grow on without needing a large reactor to hold them, and because the nutrient reaching the cells can be kept fresh. The reactor also provides an easy way of removing the product that the cells are making: such as monoclonal antibodies. Hollow fibre reactors are less use when the cells themselves have to grow, because it is hard to get at the inside of the fibre to remove surplus cells.

■ **holometabolous**

an insect that undergoes complete metamorphosis to the adult from a morphologically distinct larval stage.

■ **homeobox**

a DNA sequence found in several genes that are involved in the specification of organs in different body parts in animals; they are characteristic of genes that influence segmentation in animals. The homeobox corresponds to an amino acid sequence in the polypeptide encoded by these genes; this sequence is called the homeodomain.

■ **homeotic mutation**

a mutation that causes a body part to develop in an inappropriate position in an organism, such as the mutation in *Drosophila* that causes legs to develop on the head in place of antennae.

■ **homoalleles**

mutations that are both functionally and structurally allelic; mutations at the same site in the same gene.

■ **homodimer**

a protein with two identical polypeptide chains.

■ homogametic sex

producing similar gametes with regard to the sex chromosomes. In mammals, the XY male is heterogametic, and the XX female is homogametic.

■ homogenotisation

a genetic technique used to replace one copy of a gene, or other DNA sequence within a genome, with an altered copy of that sequence. The DNA is first cloned and then altered in some way, e.g., a transposon is inserted into a gene. The mutated gene copy can be used to replace the original gene by recombination *in vivo*. The incorporation of the mutated gene is usually selected, for example, by virtue of its containing a transposon-encoded antibiotic resistance.

■ homokaryon

cell with two or more identical nuclei as a result of fusion. Opposite: heterokaryon.

■ homologous

from the same source, or having the same evolutionary function or structure.

■ homologous chromosomes

chromosomes that occur in pairs and are generally similar in size and shape: one comes from the male parent and the other from the female. Such chromosomes contain the same linear array of genes.

■ homologous recombination

the exchange of DNA fragments between two DNA molecules or chromatids of paired chromosomes (during crossing over) at the site of identical nucleotide sequences.

■ homology

the degree of identity between individuals, or characters. The degree of identity between the nucleotide sequences of two nucleic acid molecules or the amino acid sequences of two protein molecules. Although sequence determination is the ultimate test of homology, useful estimates can be provided by either DNA-DNA or DNA-RNA hybridisation.

■ **homoplasmy**
a cellular condition in which all copies of an organelle are genetically identical. *See* heteroplasmy.

■ **homopolymer**
a nucleic acid strand that is composed of one kind of nucleotide, e.g., GGGGGGG.

■ **homozygote**
(Gr. *homos*, one and the same + *zygon*, yoke) an individual that has two copies of the same allele at a particular locus in its two homologous chromosomes. *See* allele; genotype; heterozygote.

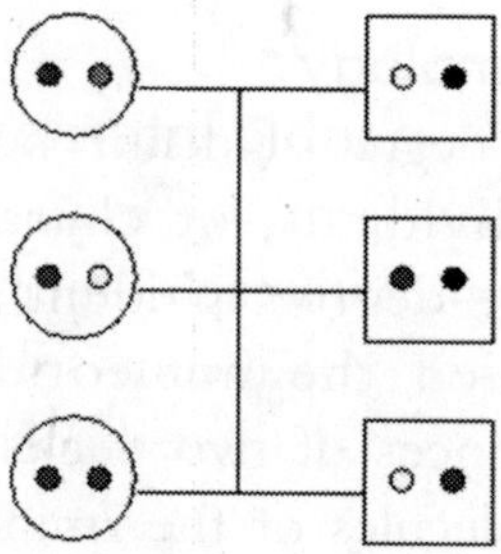

■ **hormone**
a specific organic product, produced in one part of a plant or animal body, and transported to another part where, at low concentrations, it promotes, inhibits or quantitatively modifies a biological process.

■ **host**
an organism that contains another organism, or a cloning vector.

■ **host-specific toxin**
a metabolite produced by a pathogen which has a host specificity equivalent to that of the pathogen. Such toxins are utilized for *in vitro* selection experiments to screen for tolerance or resistance to the pathogen.

■ **humoral immune response**
the production of antibody by B cells of the immune system in response to the presence of a foreign antigen. *See* antibody-mediated immune response.

■ **hybrid selection**
the process of choosing individuals possessing desired characteristics from among a hybrid population.

■ **hybrid vigour**
the extent to which the performance of a hybrid in one or more traits is better than the average performance of the two parental populations.

■ **hybridisation**

interbreeding of species, races, varieties and so on, among plants or animals; a process of forming a hybrid by cross pollination of plants or by mating animals of different types.

■ **hybridoma**

a hybrid cell, derived from a B (antibody producing) lymphocyte fused to a tumour cell, which grows indefinitely in tissue culture and is selected for the secretion of the specific antibody produced by that B cell.

■ **hydrate**

a compound formed by the incorporation of water.

■ **hydrogen bond**

a relatively weak bond formed between a hydrogen atom (which is covalently bound to a nitrogen or oxygen atom) and a nitrogen or oxygen atom with an unshared electron pair. Weak interactions between electro-negative atoms and hydrogen atoms (electro-positive) that are linked to other electro-negative atoms.

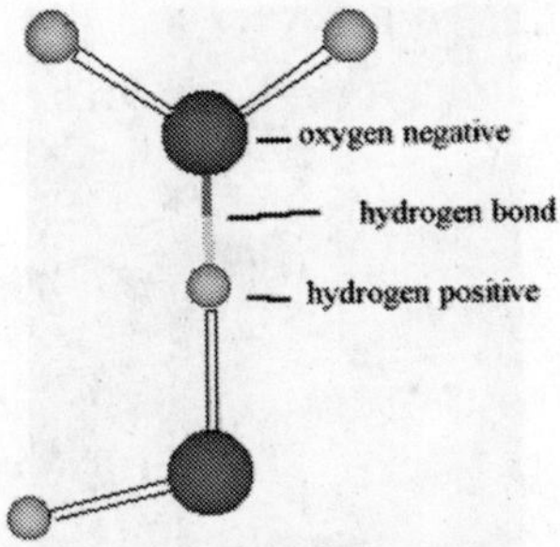

■ **hydrogen-uptake positive**

a term describing a micro-organism that is capable of assimilating (or taking up) hydrogen gas.

■ **hydrolysis**

a reaction in which a molecule of water is added at the site of cleavage of a molecule to two products.

■ **hydrophobic interactions**

association of non-polar groups with each other when present in aqueous solutions because of their insolubility in water.

■ **hydroponics**

the growing of plants in aerated water containing all the essential mineral nutrients, with no soil. Also called soilless gardening or cultivation.

■ **hydroxyl end, 3**
The hydroxyl group that is attached to the 3´ carbon atom of the sugar (ribose or deoxyribose) of the terminal nucleotide of a nucleic acid molecule.

■ **hyperploid**
a genetic condition in which a chromosome or a segment of a chromosome is over-represented in the genotype. Opposite: hypoploid.

■ **hypersensitive sites**
regions in the DNA that are highly susceptible to digestion with endonucleases.

■ **hypertonic**
a solution with an osmotic potential greater than that of living cells, leading to water loss from, shrinkage or plasmolysis of cells in a hypertonic situation.

■ **hypervariable segment**
a region of a protein that varies considerably between strains or individuals.

■ **hypochlorite**
generic term for aqueous solutions of sodium hypochlorite, potassium hypochlorite or calcium hypochlorite, which are oxidising agents and used for disinfecting surfaces and surface-sterilising tissues, and for bleaching.

■ **immediate early gene**
a viral gene that is expressed promptly after infection.

■ **immobilised cells**
cells entrapped in matrices such as alginate, polyacrylamide and agarose designed for use in membrane and filter bioreactors.

■ **immortalisation**
the genetic transformation of a cell type into a cell line which can proliferate indefinitely.

■ **immortalising oncogene**
a gene that upon transfection enables a primary cell to grow indefinitely in culture.

■ **immune response**
the processes, including the synthesis of antibodies, that are used by vertebrates to respond to the presence of a foreign antigen.

■ **immunisation**
the production of immunity in an individual by artificial means. Active immunization involves the introduction, either orally or by infection, of specially treated bacteria, viruses or their toxins so as to stimulate the production of antibodies.

■ **immunity**
the state of relative insusceptibility of an animal or plant to infection by disease-producing organisms or to the harmful effects of their poisons.

■ **immuno-affinity chromatography**
a purification technique in which an antibody is bound to a matrix and is subsequently used to bind and separate a protein from a complex mixture. *cf* affinity chromatography.

■ **immunoassay**
an assay system which detects proteins by using an antibody specific to that protein. A positive result is seen as a precipitate of an antibody-protein complex. The antibody can be linked to a radioactive atom or to an enzyme which catalyses an easily monitored reaction such as a change in colour.

■ **immunochemical control**
use of immune agents to combat infections.

■ **immunogenicity**
the ability to elicit an immune response.

■ **immunoglobulin**
one of a group of proteins (globulins) in the body that act as antibodies. They are produced by specialised cells (B lymphocytes) and are present in blood serum and other body fluids.

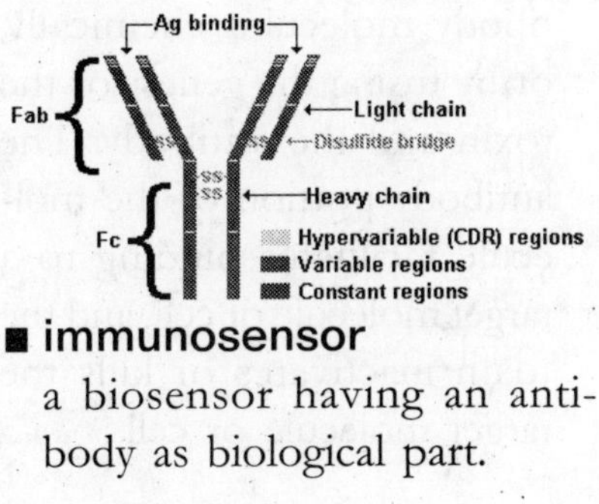

■ **immunosensor**
a biosensor having an antibody as biological part.

■ immunosuppression
the suppression of immune response. Immunosuppression is necessary following organ transplants in order to prevent the host rejecting the grafted organ.

■ immunosuppressor
a substance, an agent or a condition that prevents or diminishes the immune response.

■ immunotherapy
the use of an antibody or a fusion protein containing the antigen binding site of an antibody to cure a disease or enhance the well-being of a patient.

■ immunotoxin
protein drugs consisting of an antibody joined to a toxin molecule. Immunotoxins can be made by linking toxin and antibody molecules chemically, or by fusing the genes for the toxin and the antibody. The antibody portion of the molecule facilitates binding to a target molecule or cell, and the toxin inactivates or kills the target molecule or cell.

■ impeller
an agitator that is used for mixing the contents of a bioreactor.

■ in vivo gene therapy
the delivery of a gene or genes to a tissue or organ of an individual to alleviate a genetic disorder.

■ inactivated agent
a virus, bacterium or other organism that has been treated to prevent it from causing a disease.

■ inbred line
the product of inbreeding, i.e., the mating of individuals that have ancestors in common; in plants and laboratory animals, it refers to populations resulting from at least 6 generations of selfing or 20 generations of brother-sister mating, that are for all practical purposes, completely homozygous. In farm animals, the term is sometimes used to describe populations that have resulted from several generations of the mating of close relatives, without having reached complete homozygosity.

■ inbreeding

matings between individuals that have one or more ancestors in common, the extreme condition being self-fertilisation, which occurs in many plants and some primitive animals.

■ inbreeding depression

reduction in vigour, yield, etc., of a population that is commonly seen as the level of inbreeding increases. The traits that show greatest inbreeding depression are those that are most closely associated with viability and reproductive ability.

■ inclusion body

protein that is overproduced in a recombinant bacterium and forms a crystalline array inside the bacterial cell.

■ incompatibility

selectively-restricted mating competence, which limits fertilisation, such as lack of proper functions by otherwise normal pollen grains or certain pistils, a condition that may be caused by a variety of factors.

■ incompatibility group

a classification scheme indicating which plasmids can co-exist within a single cell. Plasmids must belong to different incompatibility groups to co-exist within the same cell. Plasmids that belong to the same incompatibility group are unstable when placed in the same cell. A plasmid cloning vector should always belong to an incompatibility group different from that of the host bacterium's endogenous plasmids.

■ incomplete dominance

a type of gene action in which heterozygotes have a phenotype that is distinctly different from, and intermediate to, the homozygous phenotypes..

■ incomplete penetrance

when some individuals in a population have a specific genotype that causes an abnormality but are not affected.

■ incubation

the hatching of eggs by means of heat, either natural or artificial.

■ incubator
an apparatus in which environmental conditions (light, photoperiod, temperature, humidity, etc.) are fully controlled, and used for hatching eggs, multiplying micro-organisms, culturing plants, etc. *cf* culture room; growth cabinet.

■ indehiscent
describing a fruit or fruiting body that does not open to release its seeds or spores when ripe.

■ independent assortment
the random distribution of alleles (from different loci) to the gametes that occurs when genes are located in different chromosomes or far apart on large chromosomes. The distribution of alleles at one locus is independent of other the distribution of alleles at another locus. *cf* linkage.

■ inheritance
the transmission of particular characteristics and/or genes from generation to generation.

■ inhibitor
any substance or object that retards a chemical reaction; a major or modifier gene that interferes with a reaction or with the expression of another gene.

■ initial
noun: cells in a meristem that remain permanently meristematic and form tissues of particular structure and function.

■ inositol lipid
a membrane-anchored phospholipid that transduces hormonal signals by stimulating the release of any of several chemical messengers. *See* phospholipid.

■ insecticide
a substance that kills insects.

■ insert
a DNA molecule that is incorporated into a cloning vector.

■ insertion element
generic term for DNA sequences found in bacteria capable of genome insertion. Postulated to be responsible for site-specific phage and plasmid integration.

■ insertion mutations
changes in the base sequence of a DNA molecule resulting from the random integration of DNA from another source.

■ **insertion site**
a unique restriction site in a vector DNA molecule into which foreign DNA can be inserted. The term is also used to describe the position of integration of a transposon or insertion site element.

■ **intercalary growth**
a pattern of stem elongation typical of grasses. Elongation proceeds from the lower internodes to the upper internodes through the differentiation of meristematic tissue at the base of each internode.

■ **intercalating agent**
a chemical capable of inserting between adjacent base pairs in a DNA molecule.

■ **intercellular space**
pore space between cells, especially typical of leaf tissues.

■ **interfascicular cambium**
cambium that arises between vascular bundles.

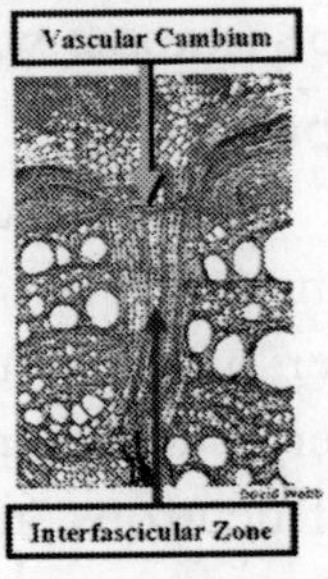

■ **interference**
crossing over at one point that alters the chance of another crossing-over nearby; detected by studying the pattern of crossings-over with three or more linked genes. Interference is positive or negative depending on whether the chance of another crossing-over nearby is reduced or increased, respectively.

■ **interferon**
a family of small proteins that stimulate viral resistance in cells.

■ **intergeneric**
a cross between two different genera.

■ **intergenic regions**
DNA sequences located between genes; they comprise a large percentage of a genome and have no known function.

■ **interleukin**
a group of proteins that transmit signals between immune cells and are necessary for mounting normal immune responses.

■ **internode**
the region of a stem between two successive nodes.

interphase
the stage in the cell cycle when the cell is not dividing; the stage follows telophase of one division and extends to the beginning of prophase in the next division. DNA replication occurs during this stage.

intersex
an organism displaying a mixture of male and female attributes.

interspecific
between two different species, e.g., an interspecific cross is a cross between two species.

isoform
a member of a family of closely related proteins that have some amino acid sequences in common and some different.

isogamy
sexual reproduction involving the production fusion of gametes that are similar in size and structure.

isogenic stocks
strains of organisms that are genetically identical; completely homozygous.

isolating mechanism
any of the biological properties of organisms that prevent interbreeding (and therefore exchange of genetic material) between members of different species that inhabit the same geographical area.

isolation medium
an optimum medium suitable for explant survival, growth and development.

isoleucine
see **amino acid.**

isomerase
any of a class of enzymes that catalyse the re-arrangement of the atoms within a molecule, thereby converting one isomer into another.

isotonic
solutions having the same osmotic potential; the same molar concentration. For protoplasts to survive, the medium they are suspended in must be isotonic with them.

isotope
one of two or more forms of an element that have the same number of protons (atomic number) but differing numbers of neutrons (mass num-

bers). Radioactive isotopes are commonly used to make DNA probes and metabolic tracers.

■ isozyme

a variant of a particular enzyme. In general, all the isozymes of a particular enzyme have the same function and sometimes the same activity, but they differ in amino-acid sequence. With the help of isozyme analysis, based on electrophoresis techniques that can separate the different variants, related species or cultivars can be distinguished.

■ kinetochores

two parallel bushlike filaments situated in the centromere of a chromosome.

■ kinetosome

granular cytoplasmic structure which forms the base of a cilium or flagellum. *cf* basal body.

■ kinin

a substance promoting cell division. In plant systems, the prefix cyto- has been added (cytokinin) to distinguish it from kinin in animal systems.

■ Klenow fragment

a product of proteolytic digestion of the DNA polymerase I from *E. coli*; it has both polymerase and 3′-exonuclease activities but not 5′-exonuclease activity.

■ Km

a dissociation constant that characterises the binding of an enzyme to a substrate. The smaller the value of K_m, the tighter the binding of the enzyme to the substrate. Also called the Michaelis constant.

■ knockout

an animal resulting from an embryonic stem cell in which a normal functional gene has been replaced by a non-functional form of the gene. This technique is used extensively in mice - much can be learned about the function of a gene by studying the phenotype of animals that lack the peptide product of that gene.

■ label

a compound or atom that is either attached to or incorporated into a macromolecule and is used to detect the presence of a compound, substance, or macromolecule in a sample. a.k.a. tag.

■ **labelling**

the process of replacing a stable atom in a compound with a radioactive isotope of the same element to enable it to be detected by autoradiography or other techniques. Increasingly, radioactive labelling is being replaced by fluorescent labelling. The method is used to trace the path of the labelled compound through a biological or chemical system.

■ **lactose**

milk sugar, a disaccharide with one unit each of glucose and galactose.

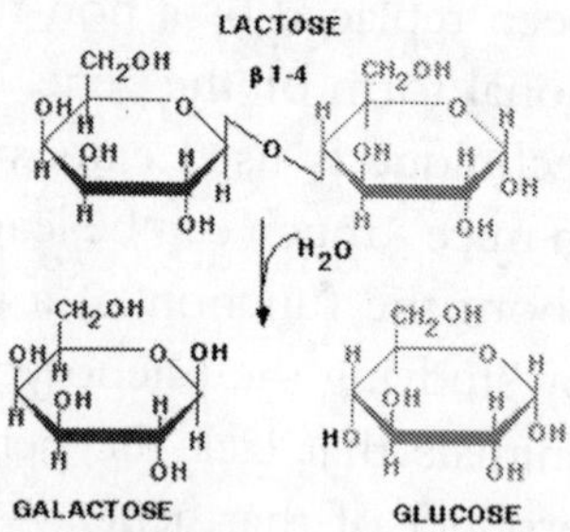

■ **lag phase**

a state of apparent inactivity preceding a response; called also a latent phase.

■ **lagging strand**

the strand of DNA that is synthesised discontinuously during replication (because DNA synthesis can proceed only in the 5′ to 3′ direction).

■ **lambda chain**

one of two classes of antibody light chains. *cf* kappa chain.

■ **lamella**

a double-membrane structure, plate or vesicle that is formed by two membranes lying parallel to each other.

■ **lamina**

blade or expanded part of a leaf.

■ **laminar air-flow cabinet**

cabinet for inoculation of cultures. The working area is kept sterile by a continuous, non-turbulent flow of sterilised air through a HEPA filter. *See* HEPA filter; prefilter.

■ **laminarin**

a storage polysaccharide of the brown algae.

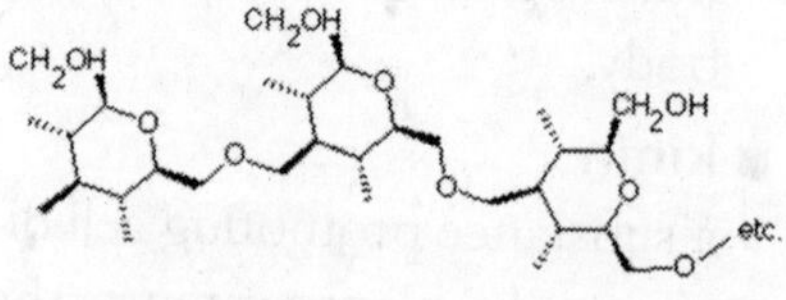

■ **lampbrush chromosomes**

large diplotene chromosomes present in oocyte nuclei, and

particularly conspicuous in amphibians. These chromosomes have extended regions called loops, which are active sites of transcription.

■ **landrace**
in PGR: an early, cultivated form of a crop species, evolved from a wild population.

■ **latent agent**
something, usually a virus, that is present in a host organism without producing any symptoms.

■ **latent bud**
an inactive bud not held back by rest or dormant period, but which may start growth if stimulated.

■ **life cycle**
the complete sequence of events undergone by organisms of a particular species, from the fusion of gametes in one generation to the same stage in the following generation.

■ **ligand**
a molecule that can bind to another molecule in or on cells.

■ **lipofection**
delivery into eukaryotic cells of DNA, RNA or other compounds that have been encapsulated in an artificial phospholipid vesicle.

■ **lipopolysaccharide**
a compound containing lipid bound to a polysaccharide; often a component of microbial cells walls.

■ **liposome**
a microscopic artificial membrane vesicle consisting of a spherical phospholipid bilayer. Liposomes can be incorporated into living cells and used to transport relatively toxic drugs into diseased cells, where they can exert their maximum effect. DNA molecules may be entrapped in, or bound to the surface of, the vesicles, and subsequent fusion of the liposome with the cell membrane will deliver the DNA into the cell. Liposomes have been used to develop an efficient transfection procedure for *Streptomyces* bacteria.

Liposomes

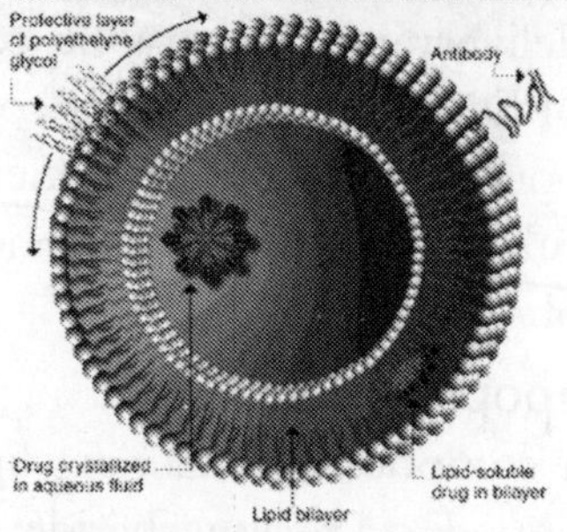

■ liquefaction

enzymatic digestion (often by a-amylase) of gelatinised starch to form lower molecular weight polysaccharides.

■ liquid media

media without a solidifying agent.

■ liquid membranes

thin films made up of liquids (as opposed to solids) which are stable in another liquid (usually water). Thus the liquid must not dissolve in the water, but nevertheless must be prevented from collapsing into a lot of small droplets.

■ liquid nitrogen

nitrogen gas condensed to a liquid with a boiling point of about -196°C. Very commonly the medium in which containers of genetic material are stored.

■ litmus

a pH indicator paper (range 4.5-8.3) impregnated with an extracted lichen pigment. It turns red in acidic and blue in alkaline solutions. However, the use of litmus paper as an indicator is not a precise method of pH measurement.

■ live vaccine

a living, non-virulent form of a micro-organism or virus that is used to elicit an antibody response that will protect the inoculated organism against infection by a virulent form of the micro-organism or virus. Also a living, non-virulent micro-organism or virus that express a foreign antigenic protein and is used to inoculate humans or animals. The latter organisms are also called a live recombinant vaccines.

■ locus

a site on a chromosome.

■ lod score

the logarithm of the odds of linkage between two loci. Calculated from pedigree data, as the log (to base 10) of the ratio of the probability of the observed pedigree assuming linkage with a specified recombination fraction q, to the probability of the observed pedigree assuming no linkage, i.e., recombination fraction = 0.5. A lod score (also called a *z* value) is thus calculated as $z = \log_{10} \{\text{Prob(data} | \text{q)} / \text{Prob(data} | 0.5)\}$. A lod score of +3.00 (which is odds of 1000:1) or greater is regarded as acceptable evidence for linkage; -2.00 (which is the log of 1:100) or less indicates that no linkage exists.

■ logarithmic phase
the steepest slope of the growth curve; the phase of vigorous growth, during which cell number doubles every 20-30 minutes..

■ long template
a DNA strand that is synthesised during the polymerase chain reaction and has a primer sequence at one end but is extended beyond the site that is complementary to the second primer at the other end.

■ long terminal repeat
a string of bases that occurs at each end of the "genome" of a retrovirus that has become integrated into the host genome. Involved in the integration process.

■ long-day plant
plant requiring short nights before flowering is initiated.

■ loop bioreactors
fermenters in which the fermenting material is cycled between a bulk tank and a smaller tank or loop of pipes. The circulation helps to mix the materials and to ensure that gas injected into the fermenter is well distributed in the liquid. The reactors are also very useful for photosynthetic fermentations, where they allow the photosynthesizing organisms to be passed along a large number of small pipes, where the light can get to them easily, rather than inside a single volume, where only the organisms near the edges get much light.

■ luteinising hormone
a pituitary hormone which causes growth of the yellow

body of the ovary and also stimulates activity of the interstitial cells of the testis.

■ **lux**

the unit of measurement for illuminance (i.e., the amount of illumination) impinging upon a surface. 1 lx is the illuminance impinging upon a surface of 1 m^2, each point of which is at a distance of 1 m away from a uniform point source of light of 1 cd (candela). It supersedes the foot-candle. *See* foot-candle; photon.

■ **luxury consumption**

nutrient absorption by an organism in excess of that required for optimum growth and productivity.

■ **lyase**

any of a class of enzymes that catalyse either the cleavage of a double bond and the addition of new groups to a substrate, or the formation of a double bond.

■ **lymphocyte**

a general class of white blood cells that are important components of the immune system of vertebrates.

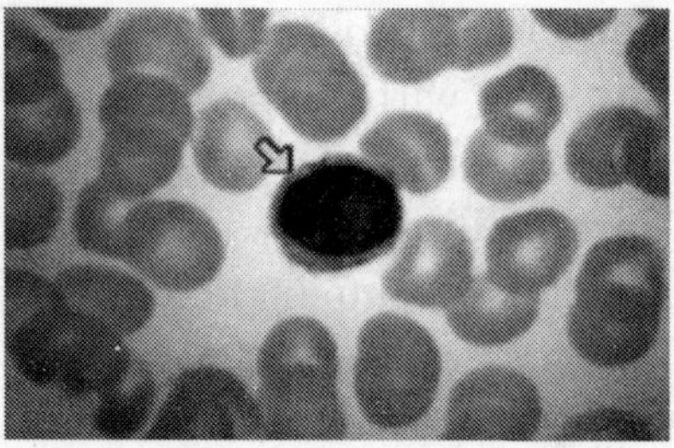

■ **lymphokine**

generic name for proteins that are released by lymphocytes to act on other cells involved in the immune response. The term includes interleukins and interferons.

■ **lymphoma**

cancer originating in the lymph nodes, spleen and other lympho-reticular sites.

■ **lyophilise**

rapid freezing followed by a dehydration under high vacuum. The process is lyophilization.

■ **lysogen**

a bacterial cell whose chromosome contains integrated viral DNA.

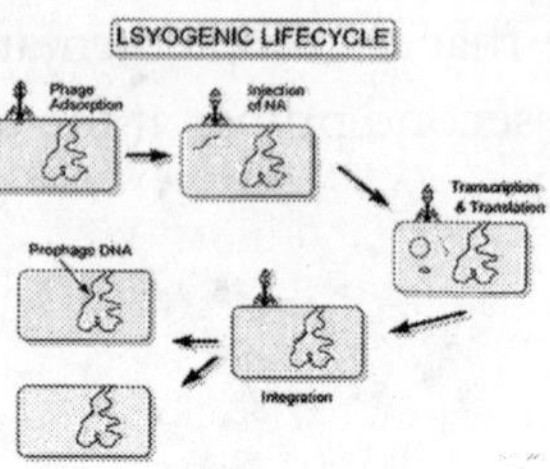

■ lysogenic

a type or phase of the virus life cycle during which the virus integrates into the host chromosome of the infected cell, remaining essentially dormant for some period of time.

■ lysogenic bacteria

bacteria harbouring temperate (non-virulent, lysogenic) bacteriophages.

■ lysogeny

a condition in which a bacteriophage genome (pro-phage) survives within a host bacterium, either as part of the host chromosome or as part of an extrachromosomal element, and does not initiate lytic functions.

■ lysosome

a membrane-bound sac within the cytoplasm of animal cells that contains enzymes responsible for the digestion of material in food vacuoles, the dissolution of foreign particles entering the cell and, on the death of the cell, the breaking down of all cell structures. The digestive system of the cell.

■ lytic

a phase of the virus life cycle during which the virus replicates within the host cell, releasing a new generation of viruses when the infected cell lyses.

■ lytic cycle

the steps in viral production that usually lead to cell lysis.

■ M13 strand

the single-stranded DNA molecule that is present in the infective form of bacteriophage M13.

■ macerate

to disintegrate tissues to obtain a cell dissociation. Cutting, soaking or enzymatic actions are commonly used.

■ macromolecule

molecule of large molecular weight, such as proteins, nucleic acids and polysaccharides.

■ map distance

the standard measure of distance between loci, expressed in centiMorgans (cM). Estimated from recombination fraction via a mapping function (q.v.). For small recombination fractions, map distance equals the percentage of recombination (recombination frequency) betwen two genes. 1% recombination = 1 cM. Sometimes called a map unit.

■ mapping

determining the location of a locus (gene or genetic marker) on a chromosome. *See* continuous map; linkage map; physical map.

■ mean

in statistics, the arithmetic average; the sum of all measurements or values in a sample divided by the sample size.

■ median

in a set of measurements, the central value above and below which there are an equal number of measurements.

■ meristem culture

a tissue culture containing meristematic dome tissue without adjacent leaf primordia or stem tissue. The term may also imply the culture of meristemoidal regions of plants, or meristematic growth in culture.

■ meristem tip

an explant comprising the meristem (meristematic dome) and usually one pair of leaf primordia. Also refers to explants originating from apical meristem tip or lateral or axillary meristem tip. Do not confuse the meristem tip with the term "shoot tip," which is much larger and usually has more immature leaves and stem tissue.

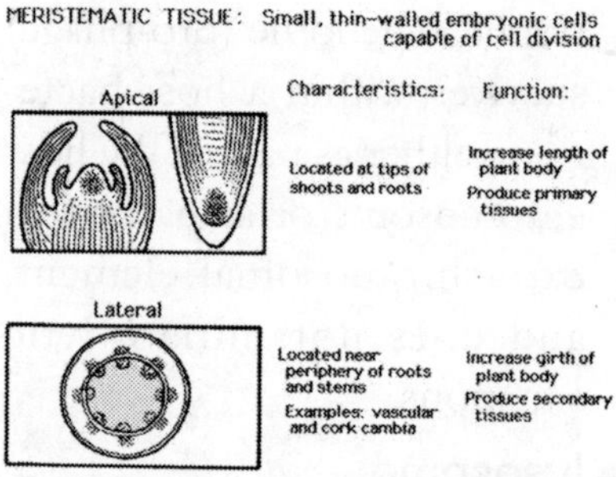

■ meristem tip culture

cultures derived from meristem tip explants. The use of meristem tip culture is for virus elimination or axillary shoot proliferation purposes, but less commonly for callus production.

■ meristemoid

a localised group of cells in callus tissue, characterised by an accumulation of starch, RNA and protein, and giving rise to adventitious shoots or roots.

■ merozygote
partial zygote produced by a process of partial genetic exchange, such as transformation in bacteria.

■ mesoderm
the middle germ layer that forms in the early animal embryo and gives rise to parts such as bone and connective tissue.

■ mesophile
a micro-organism able to grow in the temperature range 20 to 50°C; optimal growth often occurs at about 37°C.

■ mesophyll
leaf parenchyma tissue occurring between epidermal layers.

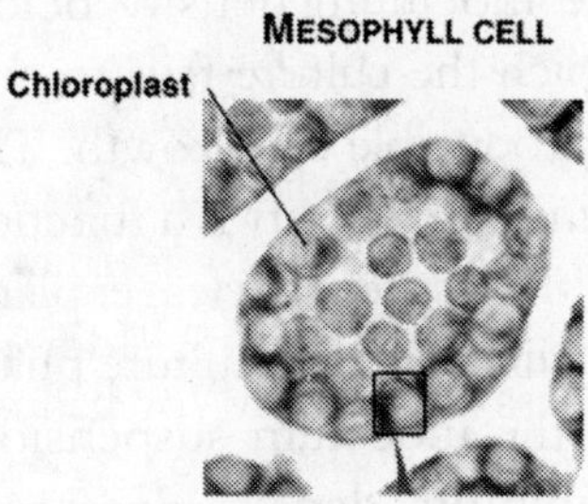

■ messenger RNA
see mRNA.

■ metabolic cell
a cell that is not dividing.

■ metabolism
in an organism or a single cell, the biochemical process by which nutritive material is built up into living matter, or aids in building living matter, or by which complex substances and food are broken down into simple substances.

■ metabolite
a low-molecular-weight biological compound that is usually synthesised by an enzyme.

■ micro-element
an element required in very small quantities.

■ micro-encapsulation
a process of enclosing a substance in very small sealed capsules from which material is released by heat, solution or other means.

■ micro-environment
the environment close enough to the surface of a living or non-living object to be influenced by it.

■ microfibrils
microfibrils are exceedingly small fibres visible only at the high magnification of the electron microscope.

■ micro-injection
the introduction of small amounts of material (DNA, RNA, enzymes, cytotoxic agents) into a single eukaryotic cell with a fine, microscopic needle, penetrating the cell membrane.

■ micro-isolating system
mechanical separation of single cells or protoplasts thus allowing them to proliferate individually.

■ micron
a unit of distance: 10^{-6} m; 0.001 mm. Symbol: mm.

■ micronutrient
for growth media: An essential element normally required in concentrations < 0.5 millimole/litre.

■ micro-organism
organism visible only under magnification.

■ microplasts
vesicles produced by subdivision and fragmentation of protoplasts or thin-walled cells.

■ micropyle
a small opening in the surface of a plant ovule through which the pollen tube passes prior to fertilisation.

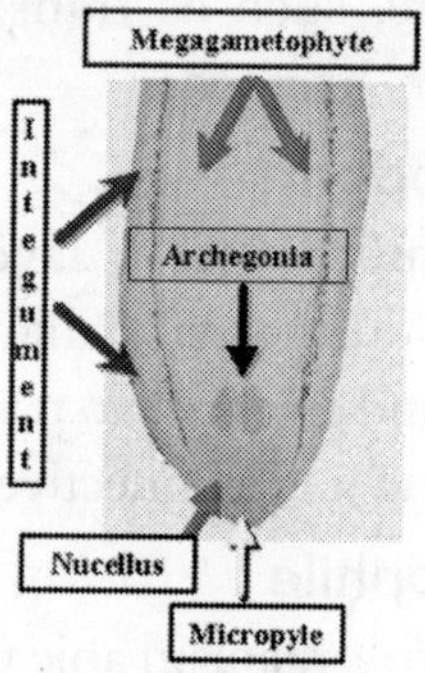

■ mid-parent value
in quantitative genetics, the average of the phenotypes of two mates.

■ minimum effective cell density
the inoculum density below which the culture fails to give reproducible cell growth. The minimum density is a function of the tissue (species, explant, cell line) and the culture phase of the inoculum suspension. Minimum density decreases inversely to the aggregate size and division rate of the stock culture.

■ minimum inoculum size
the critical volume of inoculum (q.v.) required to initiate culture growth, due to the diffusive loss of cell materials into the medium. The subsequent culture growth cycle is dependent on the inoculum size, which is determined by the volume of medium and size of the culture vessel.

■ mini-prep
a small-scale (mini-) preparation of plasmid or phage DNA. Used to analyse DNA in a cloning vector after a cloning experiment.

■ minisatellite
a form of VNTR (q.v.) in which the repeat units (q.v.) typically range from 10 to 100 bases. They are usually detected by Southern hybridisation (q.v.), using a probe comprising a clone of the repeat unit. The first DNA fingerprints (q.v.) were minisatellites detected in this way. Minisatellites tend to be located at the ends of chromosomes and in regions with a high frequency of recombination.

■ minitubers
small tubers (5-15 mm in diameter) formed on shoot cultures or cuttings of tuber-forming crops, such as potato.

■ mismatch
the lack of a complementary pair of bases in a double helix of DNA, e.g., A:C, G:T.

■ mismatch repair
DNA repair processes that correct mismatched base pairs.

■ missense mutation
a mutation that changes a codon for one amino acid into a codon specifying another amino acid.

■ mist propagation
application of fine droplets of water to leafy cuttings in the rooting stage to reduce transpiration. *cf* fog.

■ mites
free-living and parasitic animals belonging to the order Acarina, class Arachnida (with spiders). Mites may infest plant crops, reducing their harvest. They may also infest plant tissue culture work areas and incubation facilities in search of sugars, and

so contaminate culture vessels and spread bacteria and fungi.

■ **mitochondrial DNA**

a circular ring of DNA found in mitochondria. In mammals, mtDNA makes up less than 1% of the total cellular DNA, but in plants the amount is variable. It codes for ribosomal RNA and transfer RNA, but only some mitochondrial proteins (up to 30 proteins in animals), the nuclear DNA being required for encoding most of these.

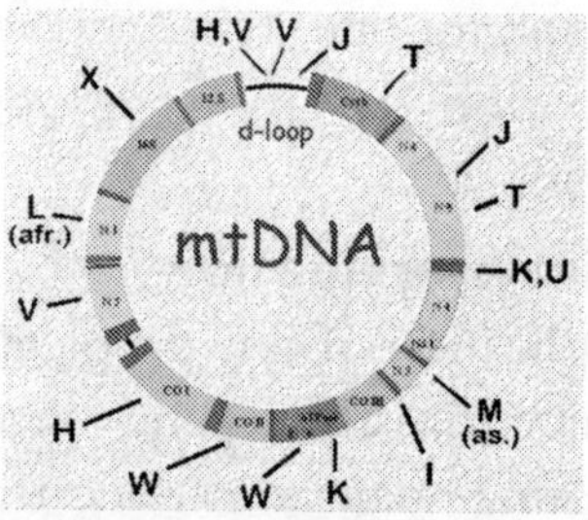

■ **mobilising functions**

the genes on a plasmid that give it the ability to facilitate the transfer of either a non-conjugative or a conjugative plasmid from one bacterium to another.

■ **modal class; mode**

in a frequency distribution, the class having the greatest frequency.

■ **model**

a mathematical description of a biological phenomenon.

■ **molecular biology**

the area of knowledge concerned with the molecular aspects of organisms and their cells.

■ **molecular cloning**

the biological amplification of a specific DNA sequence through mitotic division of a host cell into which it has been transformed or transfected. *See* cloning.

■ **molecular genetics**

the area of knowledge concerned with the genetic aspects of molecular biology, especially with DNA, RNA and protein molecules.

■ **molecule**

a unit of matter, the smallest portion of an element or a compound that retains chemical identity with the substance in mass. The molecule usually consist of a union of two or more atoms; some organic molecules containing a very large number of atoms.

■ monohybrid cross
a cross between parents differing in only one trait or in which only one trait is being considered.

■ monolayer
a single layer of cells growing on a surface.

■ monomer
a single molecular entity that may combine with others to form more complex structures.

■ monosaccharide
a single sugar. *cf* polysaccharide.

■ monosomic
describing a diploid organism lacking one chromosome (2n -1) of its proper (disomic) complement; a form of aneuploidy. *See also* disomy.

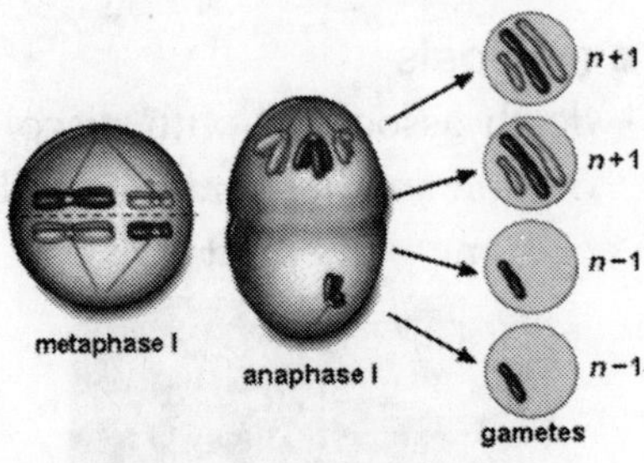

■ mono-unsaturates
oils containing mono-unsaturated fatty acids.

■ monozygotic twins
one-egg or identical twins; twins derived from the splitting of a single fertilised ovum.

■ morphogen
a substance that stimulates the development of form or structure in an organism.

■ morphogenesis
the development, through growth and differentiation, of form and structure in an organism.

■ morphogenic response
the effect on the developmental history of a plant or its parts exposed to a given set of growth conditions or to a change in the environment.

■ multivalent vaccine
a single vaccine that is designed to elicit an immune response either to more than one infectious agent or to several different epitopes of a molecule.

■ mutable genes
genes with an unusually high mutation rate.

■ mutagen
an agent or process which is capable of inducing a muta-

tion, such as UV light. *cf* mutation.

mutagenesis
change(s) in the genetic constitution of a cell through alterations to its DNA.

mutant
an organism or an allele that differs from the wild type because it carries one or more genetic changes in its DNA. A mutant organism may carry mutated gene(s) (= gene mutation); mutated chromosome(s) (= chromosome mutation); or mutated genome(s) (= genome mutation). a.k.a. a variant.

myeloma
a plasma cell cancer.

N_2
free nitrogen gas. In liquid form, used as a cryopreservant. *See* cryobiological preservation.

naked bud
a bud not protected by bud scales.

nanometre
1×10^{-9} m. One millionth of a millimetre, a.k.a. a millimicron; equals ten angstroms.

narrow-host-range plasmid
a plasmid that can replicate in one, or at most a few, different bacterial species.

narrow-sense heritability
in quantitative genetics, the proportion of the phenotypic variance that is due to variation in breeding values.

native protein
the naturally occurring form of a protein.

natural selection
the differential survival and reproduction of organisms because of differences in characteristics that affect their ability to utilize environmental resources.

NDP
Ribonucleoside diphosphate.

necrosis
death associated with discoloration and dehydration of all or some parts of organs.

■ **negative autogenous regulation**
inhibition of the expression of a gene or set of coordinately regulated genes by the product of the gene or the product of one of the genes.

■ **negative control system**
a mechanism in which the regulatory protein(s) is required to turn off gene expression.

■ **negative selection**
selection of individuals that do not possess a certain character. A method by which growing cells that do not carry a DNA insert integrated at a specific chromosomal location are selected. *See* positive selection.

■ **nematodes**
a class of slender, unsegmented worms, often parasitic. a.k.a. eelworms, especially when phytoparasitic.

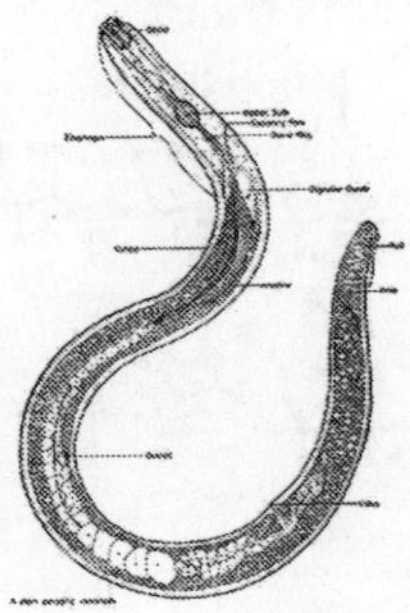

■ **neo-formation**
organogenesis; production of newly formed structures, such as tissues, meristems and embryos.

■ **neoplasm**
localised cell multiplication. Generally it designates a collection of cells which have undergone genetic transformation, forming a tumour. Neoplasmic cells differ in structure and function from the original cell type.

■ **neoteny**
the retention of juvenile body characters in the adult state, or the occurrence of adult characters in the juvenile state.

■ **net photosynthesis**
photosynthetic activity minus respiratory activity, as measured by carbon dioxide exchange.

■ **neutral mutation**
a mutation that changes the nucleotide sequence of a gene but has negligible effect on the fitness of the organism.

■ **neutral theory**
the theory that much of evolution has been primarily due

to random drift of neutral mutations.

■ **nick**

verb: to break a phospho-diester bond in the backbone of one of the strands of a duplex DNA molecule. *cf* cleave; cut.

■ **nick translation**

a procedure for labelling DNA. A DNA fragment is treated with DNase to produce single-stranded nicks. The nick is moved along the DNA molecule in the presence of labelled deoxyribonucleoside triphosphates by the concerted action of the 5´-> 3´ exonuclease and 5´-> 3´ polymerase activities of *E. coli* DNA polymerase I.

■ **nicked circle**

during extraction of plasmid DNA from the bacterial cell, one strand of the DNA becomes nicked. This relaxes the torsional strain needed to maintain supercoiling, producing the familiar form of plasmid. *See* plasmid.

■ **nitrate**

the only form in which nitrogen can be used directly by plants; a component of chemical fertilisers.

■ **nitrification**

a chemical process in which nitrogen in plant and animal wastes and dead remains is oxidised, first to nitrites and then to nitrates.

■ **nitrite**

a salt or ester of nitrous acid.

■ **nitrocellulose**

a nitrated derivative of cellulose. It is made into membrane filters of defined porosity, used to immobilise DNA, RNA or protein, which can then be probed with a labelled sequence or antibody. These filters have a variety of uses in molecular biology, particularly in nucleic acid hybridisation experiments. Used extensively in the Southern and northern blotting procedures involving DNA and RNA.

Nitrocellulose

64

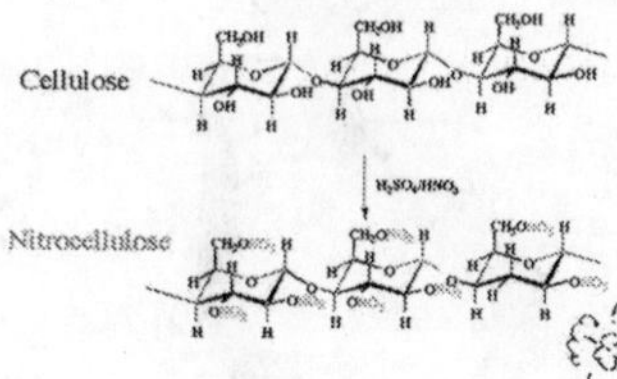

■ **nitrogen assimilation**
the incorporation of nitrogen into organic cell substances by living organisms.

■ **nitrogen fixation**
the conversion of atmospheric nitrogen (N_2) into oxidised forms that can be assimilated by plants. Biological nitrogen fixation is catalysed by the enzyme nitrogenase, which is found only in prokaryotes. Certain blue-green algae and some genera of bacteria (e.g., *Rhizobium* spp.; *Azotobacter* spp.) are capable of biochemically fixing nitrogen. Such bacteria are very important symbionts for plants growing in nitrogen-poor soils.

■ **nitrogenous bases**
the purines (adenine and guanine) and pyrimidines (thymine, cytosine and uracil) that form DNA and RNA molecules.

■ **NMP**
Ribonucleoside monophosphate.

■ **nod box**
a DNA sequence that controls the transcriptional regulation of *Rhizobium* nodulation genes.

■ **nodal culture**
the culture of a lateral bud and a section of adjacent stem tissue.

■ **node**
slightly enlarged portion of the stem where leaves and buds arise and where branches originate. Stems have nodes but roots do not.

■ **nodular**
term commonly used to describe a pebbly (rough) texture of a callus.

■ **nodulation**
the formation of nodules by symbiotic bacteria on the roots of plants.

■ **nodule**
the enlargement or swelling on roots of nitrogen-fixing plants. The nodules contain symbiotic nitrogen-fixing bacteria. *See* nitrogen fixation.

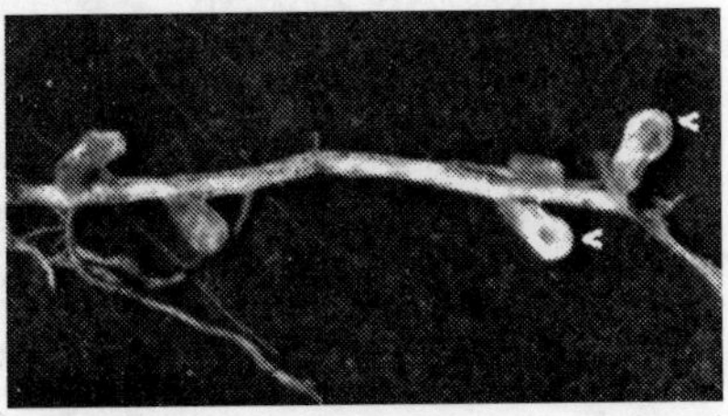

non-autonomous

a term referring to biological units that cannot function by themselves; such units require the assistance of another unit, or "helper". *See* autonomous.

non-disjunction

failure of disjunction or separation of homologous chromosomes or chromatids in mitosis or meiosis, resulting in too many chromosomes in some daughter cells and too few in others. Examples: In meiosis, both members of a pair of chromosomes go to one pole so that the other pole does not receive either of them; in mitosis, both sister chromatids go to the same pole.

non-histone chromosomal proteins

in chromosomes, all of the proteins except the histones.

nonsense mutation

a mutation which converts an amino-acid-specifying codon into a stop codon, e.g., a change from UAU (tyr) to UAG (amber) would lead to the premature termination of a polypeptide chain at the place where a tyrosine was inserted in the wild-type. *See* stop codon; suppressor.

non-target organism

an organism which is affected by an interaction for which it was not the intended recipient.

non-template strand

in transcription, the non-transcribed strand of DNA. a.k.a. sense strand or coding strand. It will have the same sequence as the RNA transcript, except that T is present at positions where U is present in the RNA transcript.

oncogenesis

the progression of cytological, genetic and cellular changes that culminate in a tumour.

oncogenic

oncogenic genes are responsible for the transformation of normal cells into tumour cells.

onco-mouse

a mouse that has been genetically modified to incorporate an oncogene; oncogenes cause cells to undergo cancerous transformation.

ontogeny

developmental life history of an organism.

■ **oocyte**

the egg mother cell; it undergoes two meiotic divisions (oogenesis) to form the egg cell. The primary oocyte is before completion of the first meiotic division; the secondary oocyte is after completion of the first meiotic division.

■ **oogenesis**

the formation and growth of the egg or ovum in an animal ovary.

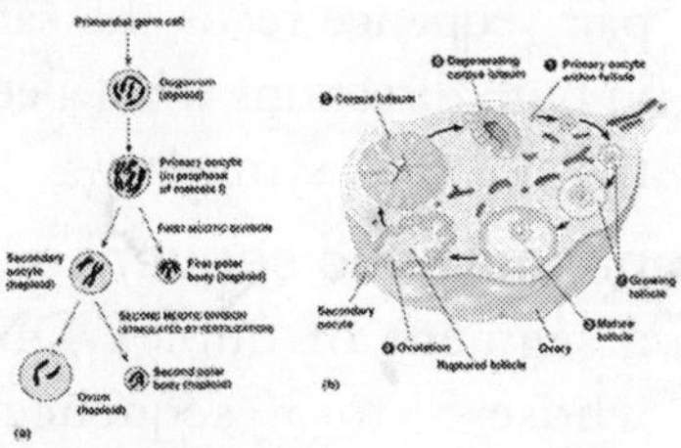

■ **oogonium**

a germ cell of the female animal, that gives rise to oocytes by mitotic division.

■ **overdominance**

a condition in which heterozygotes are superior (on some scale of measurement) to either of the associated homozygotes.

■ **overlapping reading frames**

start triplets in different reading frames generate different polypeptides from the same DNA sequence. *See* reading frame.

■ **ovulation**

in mammals, the process of escape of the ovum (egg cell) from the ovary.

■ **ovule**

the part of the reproductive organs in seed plants that consists of the nucellus, the embryo sac and integuments.

■ **ovum**

1. a gamete of female animals, produced by the ovary.
2. the oosphere in plants.

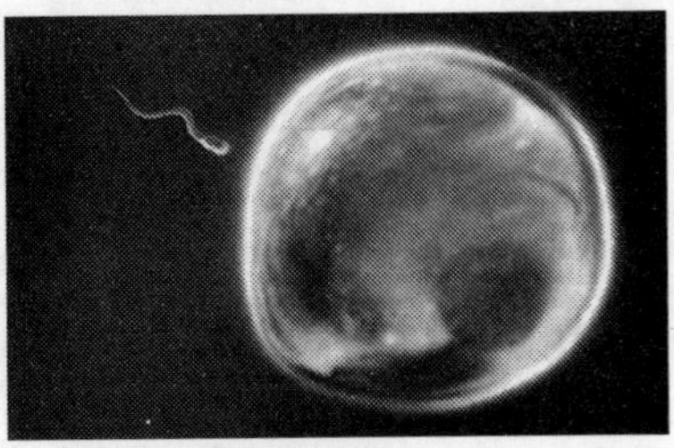

■ **ovum pickup**

the non-surgical collection of ova from a female.

■ **P_1**

symbol for the parental generation or parents of a given individual.

■ **pachynema**

a mid-prophase stage in meiosis, immediately following

zygonema and preceding diplonema. In microscopic preparations, the chromosomes are visible as long, paired threads. Rarely, four chromatids are detectable.

■ **packaging cell line**
a cell line that is designed to produce viral particles that do not contain nucleic acid. After transfection of these cells with a full-size viral genome, fully infective viral particles are assembled and released.

■ **packed cell volume**
the volume of cells in a set volume of culture expressed as a percentage of that set volume after sedimentation (packing) by means of low speed centrifugation.

■ **pairing; synapsis**
the pairing of homologous chromosomes during the prophase of the first meiotic division, when crossing over occurs.

■ **pair-rule gene**
a gene that influences the formation of body segments in *Drosophila.*

■ **palaeontology**
the study of the fossil record of past geological periods and of the phylogenetic relationships between extinct and contemporary plant and animal species.

■ **palindrome**
originally a word, sentence or verse that reads the same from right to left as it does from left to right. In biotechnology, the term is applied to a segment of DNA in which the base-pair sequence reads the same in both directions from a central point of symmetry.

■ **palindromic sequence**
a segment of duplex DNA whose 5′-to-3′ sequence is identical on each DNA strand. The sequence is the same when one strand is read left-to-right and the other strand is read right-to-left. Typically, recognition sites for type II restriction endonucleases are palindromes.

■ **palisade parenchyma**
elongated cells found just beneath the upper epidermis of leaves, and containing many chloroplasts.

■ pAMP
ampicillin-resistant plasmid.

■ panicle
an inflorescence, the main axis of which is branched; the branches bear loose racemose flower clusters.

■ panicle culture
aseptic culture of grain panicle segments to induce microspore germination and development.

■ panmictic population
a population in which mating occurs at random.

■ panmixis
random mating in a population. *cf* apomixis.

■ par gene
a gene found in bacterial and plant cells, and involved in partition of plasmids in cells.

■ paracentric inversion
an inversion that is entirely within one arm of a chromosome and does not include the centromere.

■ paraffin (wax)
a translucent, white, solid hydrocarbon with a low melting point. Paraffin is used as an embedding medium to support tissue for sectioning for light microscopy observation.

■ parafilmTM
a stretchable film based on paraffin wax; used to seal tubes and Petri dishes. ParafilmTM is a proprietary name which is applied colloquially to similar products.

■ parahormone
a substance with hormone-like properties that is not a secretory product (e.g., ethylene; carbon dioxide).

■ parallel evolution
the development of different organisms along similar evolutionary paths due to similar selection pressures acting on them.

■ parameter
a value or constant pertaining to an entire population. *cf* statistic.

■ parasexual cycle
a sexual cycle involving changes in chromosome number but differing in time and place from the usual sexual cycle; occurring in those fungi in which the normal cycle is suppressed or apparently absent.

■ parasite
an organism deriving its food from the living body of another organism.

■ parasitism
the close association of two or more dissimilar organisms, where the association is harmful to at least one. *cf* commensalism; symbiosis.

■ parasporal crystal
tightly packaged insect protoxin molecules that are produced by strains of *Bacillus thuringiensis* during the formation of resting spores.

■ parenchyma
1. a plant tissue consisting of spherical, undifferentiated cells, frequently with air spaces between them.
2. loose connective tissue formed of large cells.

■ parenchymatous
adjective used to describe spherical and undifferentiated cells with primary cell walls, capable of both cell division and differentiation.

■ parthenocarpy
the development of fruit without fertilisation.

■ parthenogenesis
production of an embryo from an unfertilised egg. *cf* androgenesis; apomixis; gynogenesis.

■ partial digest
addition of a restriction enzyme to a DNA sample under particular conditions or for a limited period, such that only a proportion of the target sites in any individual molecule are cleaved. Partial digests are often performed to give an overlapping collection of DNA fragments for use in the construction of a gene bank. *See* complete digest; library.

■ particle radiation
refers to gamma (g) particles (positively charged) and beta (ß) particles (negatively charged), electrons, protons and neutrons. In plant tissue culture, these particles are used to produce mutant cells or organisms.

■ parts per million
units of any given substance per one million equivalent units, such as the weight units of solute per million weight

units of solution (i.e., 1 ppm = 1 mg/l).

■ parturition
the process of giving birth.

■ passage
the transfer or transplantation of cells from one culture vessel to another. *cf* sub-culture.

■ passage number
the number of times the cells in the culture have been subcultured. In descriptions of this process, the dilution ratio of the cells should be stated so that the relative cultural "age" can be ascertained.

■ passage time
interval between successive sub-cultures.

■ passive immunity
1. natural acquisition of antibodies by the foetus or neonate (newborn) from the mother.
2. the artificial introduction of specific antibodies by the injection of serum from an immune animal.
in both cases, temporary protection is conferred on the recipient.

■ patent
a government-issued document that assigns the holder the exclusive right - for a defined period of time - to manufacture, use or sell an invention.

■ paternal
pertaining to the father.

■ pathogen
an organism that causes a disease in another organism.

■ pathogen-free
freedom from disease-causing organisms (bacteria, fungi, viruses, etc.).

■ pathotoxin
very dilute substance synthesised and released by some pathogen, and which interacts with the host metabolism.

■ pathovar
in plant pathology, strains of bacteria causing disease in specific plant cultivars.

■ BR322
one of the first plasmid vectors widely used; especially used for cloning DNA in *E. coli*.

■ **pectin**

a white amorphous substance which, when combined with acid and sugar, yields a jelly substance cementing cells together (the middle lamella).

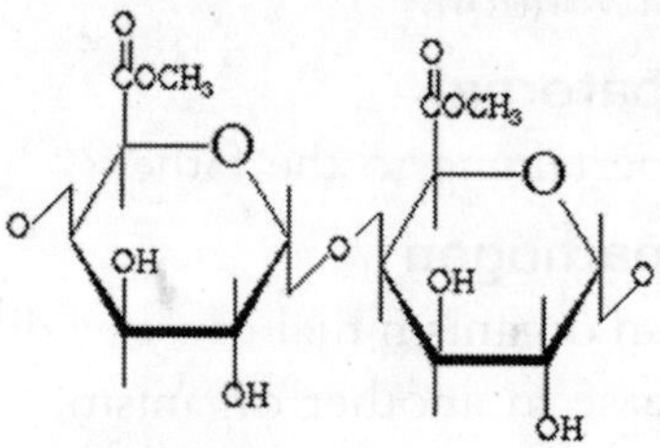

■ **pectinase**

enzyme catalysing the hydrolysis of pectins.

■ **pedicel**

stalk or stem of the individual flowers of an inflorescence.

■ **pedigree**

a table, chart or diagram recording the ancestry of an individual.

■ **peduncle**

stalk or stem of a flower that is born singly; the main stem of an inflorescence.

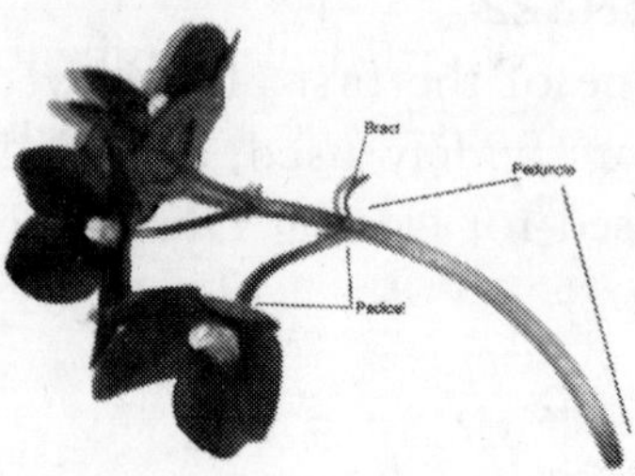

■ **penetrance**

the percentage of individuals in population that show a particular phenotype among those capable of showing it, i.e., among those that have the genotype normally associated with that phenotype.

■ **peptide**

a sequence of amino acids linked by peptide bonds; a breakdown or build-up unit in protein metabolism.

■ **peptide bond**

a chemical bond holding amino acid sub-units together in proteins.

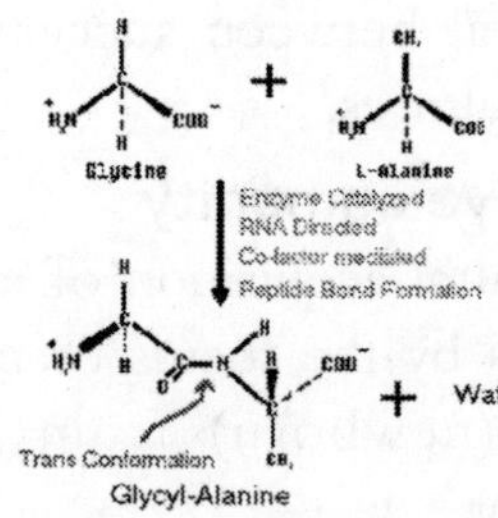

■ **peptide vaccine**

a short chain of amino acids that can induce antibodies against a specific infectious agent.

■ **peptidyl transferase**

an enzyme bound tightly to the large sub-unit of the ribosome that catalyses the formation of peptide bonds between amino acids during translation.

■ **peptidyl-tRNA binding site**
the site on a ribosome that hosts the tRNA to which an amino acid for the growing polypeptide chain is attached.

■ **perennial**
a plant that grows more or less indefinitely from year to year and, once mature, usually produces seed each year.

■ **pericentric inversion**
an inversion that includes the centromere, hence involving both arms of a chromosome.

■ **periclinal**
the plane of cell wall orientation or cell division parallel to the surface of the organ. *cf* anticlinal.

■ **periclinal chimera**
genotypically or cytoplasmically different tissues arranged in concentric layers.

■ **pericycle**
region of the plant bounded externally by the endodermis and internally by the phloem. Most roots originate from the pericycle.

■ **periplasm**
the space (periplasmic space) between the cell (cytoplasmic) membrane of a bacterium or fungus and the outer membrane or cell wall.

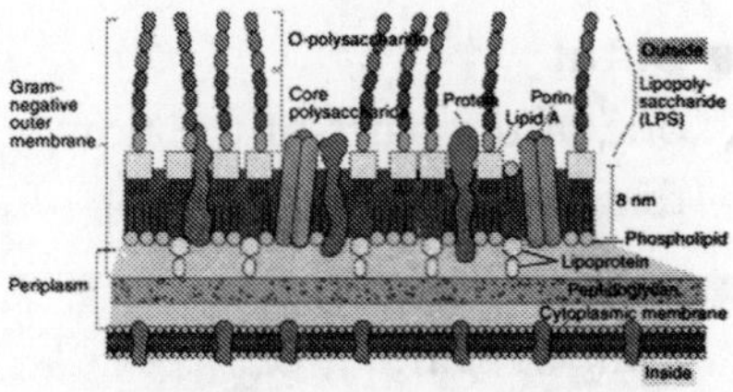

■ **permanent wilting point**
the moisture content of soil at which plants wilt to such an extent that they fail to recover even when placed in a humid atmosphere.

■ **permeable**
used of a membrane, cell or cell system through which substances may diffuse.

■ **persistence**
ability of an organism to remain in a particular setting for a period of time after it is introduced.

■ **persistent**
1.continues to exist or to remain attached.

2 chemicals with a long inactivation time, such as some pesticides, which may accumulate in the food chain.

■ **pesticide**

a toxic chemical product that kills harmful organisms (e.g., insecticides, fungicide, weedicides, rodenticides).

■ **petal**

one of the parts of the flower that make up the corolla.

■ **petiole**

stalk of leaf.

■ **petite mutant**

a respiration-deficient yeast mutant that produces small colonies when grown on glucose-containing medium.

■ **petri dish**

flat round dish with a matching lid, made of glass or plastic material, and used for culturing organisms. a.k.a. plates, hence the term to "plate" (q.v.) a culture.

■ **pH**

a measure of acidity and alkalinity. Equal to the log of the reciprocal of the hydrogen ion concentration of a solution, expressed in grams per litre. A reading of 7 is neutral (e.g., pure water), whereas below 7 is acid and above 7 is alkaline.

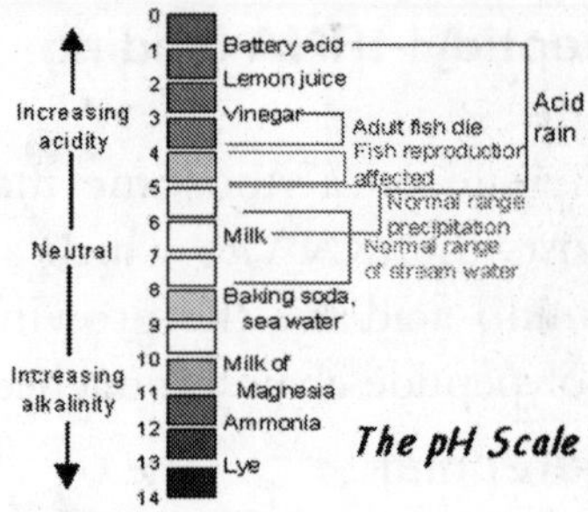

The pH Scale

■ **phagemids**

cloning vectors that contain components derived from both phage chromosomes and plasmids.

■ **phagocytes**

immune system cells that ingest and destroy viruses, bacteria, fungi and other foreign substances or cells.

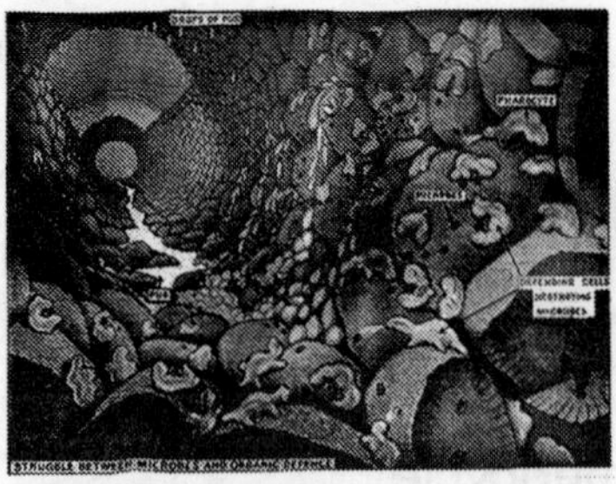

■ **phagocytosis**

the process by which minute food particles or foreign particles invading the body are engulfed and broken down by certain animal cells.

■ **phase state**
the coupling or repulsion of two linked genes.

■ **pH-electrode-based sensor**
sensor in which a standard electrochemical pH electrode is coated with a biological material. Many biological processes raise or lower pH, and the changes can be detected by the pH electrode.

■ **phenocopy**
an organism whose phenotype (but not genotype) has been changed by the environment to resemble the phenotype usually associated with a mutant organism.

■ **phenolic oxidation**
many plant species contain phenolic compounds that blacken through oxidation. The process is initiated after plants are wounded. Phenolic oxidation may lead to growth inhibition or, in severe cases, to tissue necrosis and death. Antioxidants are incorporated into the sterilising solution or isolation medium to prevent or reduce oxidative browning. *cf* activated charcoal.

■ **phenols**
compounds with hydroxyl group(s) attached to the benzene ring, forming esters, ethers and salts. Phenolic substances are produced from newly explanted tissues, oxidising to form coloured compounds visible in nutrient media.

Bonding in Phenol

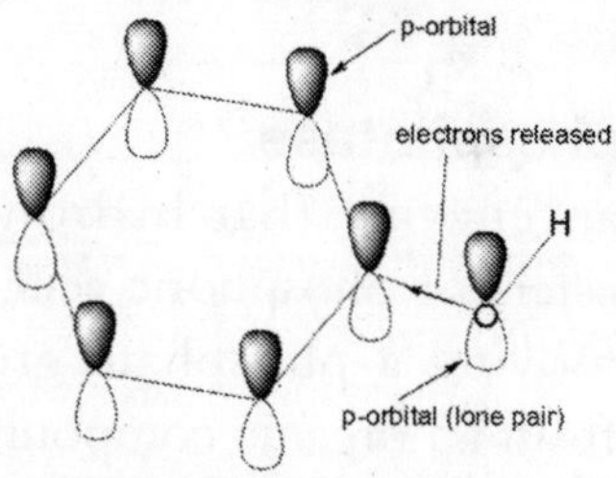

■ **phenotype**
the visible appearance or set of traits of an organism resulting from the combined action of genotype and environment. *cf* genotype.

■ **phenylalanine**
see **amino acid.**

■ **pheromone**
a hormone-like substance that is secreted by an organism into the environment as a spe-

cific signal to another organism, usually of the same species.

■ **phloem**

food-conducting tissue in plants, consisting of sieve tubes, companion cells, phloem parenchyma and fibres.

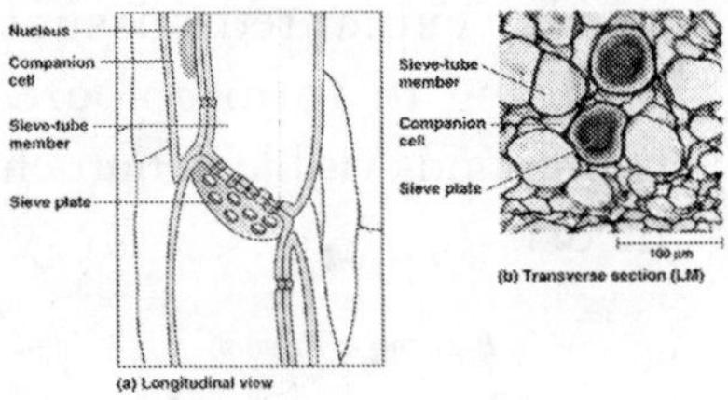

■ **phosphatase**

an enzyme that hydrolyses esters of phosphoric acid, removing a phosphate group from an organic compound.

■ **phospho-diester bond**

a bond in which a phosphate group joins adjacent carbons through ester linkages. A condensation reaction between adjacent nucleotides results in a phospho-diester bond between 3´ and 5´ carbons in DNA and RNA.

■ **phospholipase A2**

an enzyme which degrades type A2 phospholipids.

■ **phospholipid**

a class of lipid molecules in which a phosphate group is linked to glycerol and two fatty acyl groups. A major component of biological membranes. *See* inositol lipid.

■ **phosphorolysis**

the cleavage of a bond by orthophosphate; analogous to hydrolysis referring to cleavage by water.

■ **phosphorylation**

the addition of a phosphate group to a compound.

■ **photo-bioreactor**

bioreactor dependent on sunlight, which is taken up by its content of plant material, usually algae.

■ **photon**

a quantum of light; the energy of a photon is proportional to its frequency; E = hv, where E is energy; h is Planck's constant, 6.62×10^{-27} erg-second; and v is the frequency.

■ **photoperiod**

the length of day or period of daily illumination provided or required by plants for reaching the reproductive stage.

■ photoperiodism
the response of an organism to changes in day length.

■ photophosphorylation
the formation of ATP from ADP and inorganic phosphate using light energy in photosynthesis.

■ photoreactivation
a DNA repair process that is light dependent.

■ photosynthate
the carbohydrates and other compounds produced in photosynthesis.

■ photosynthesis
a chemical process by which green plants synthesise organic compounds from carbon dioxide and water in the presence of sunlight.

■ photosynthetic
able to use sunlight energy to convert atmospheric carbon dioxide into organic compounds. Nearly all plants, most algae and some bacteria are photosynthetic.

■ photosynthetic efficiency
efficiency of converting light energy into organic compounds.

■ photosynthetically active radiation
radiant energy captured by the photosynthetic system in the light reactions, usually taken to be the wavelengths between 400 and 700 nm.

■ phototropism
a growth curvature in which light is the stimulus.

■ phylogeny
a diagram illustrating the deduced evolutionary history of populations of related organisms.

■ physical map
a map showing physical locations on a DNA sequence, such as restriction sites and sequence-tagged sites. Also a diagram of a chromosome or a karyotype, showing the location of loci (genes and markers) *See* mapping.

■ phytochrome
a reversible pigment system of a protein nature, found in the cytoplasm of green plants. Phytochrome is associated with the absorption of light that affects growth, development and differen-

tiation of a plant, independent of photosynthesis, e.g., in the photoperiodic response.

■ phytohormone
a substance that stimulates growth or other processes in plants. They include auxins, abscissic acid, cytokinins, gibberellins and ethylene. Phytohormones are chemical messengers that may pass through cells, tissues and organs and stimulate biochemical, physiological and morphological responses.

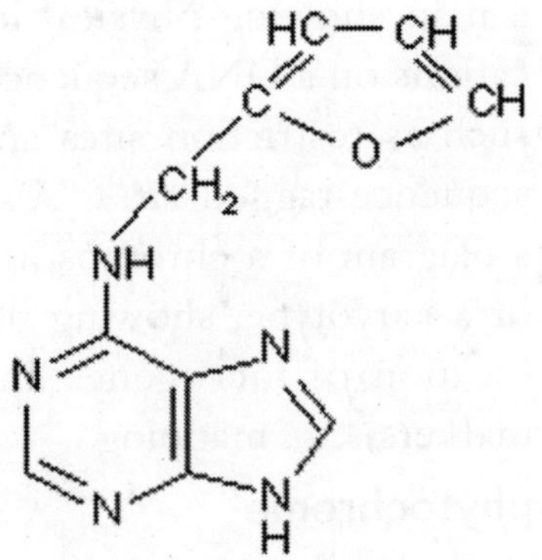

■ phytopathogen
an organism that causes disease in plants.

■ phytosanitary
plant health, including quarantine.

■ phytostat
the name adopted by Tulecke in 1965 for an apparatus designed for the semi-continuous chemostat culture of plant cells.

■ pigments
molecules that are coloured by the light they absorb. Some plant pigments are water soluble and are found mainly in the cell vacuole.

■ pinocytosis
the process by which a living cell engulfs a minute droplet of liquid.

■ pipette
a slender graduated tube into which small amounts of liquids are taken up by suction, for measuring and transferring.

■ pistil
central organ of the flower, typically consisting of ovary, style and stigma. The pistil is usually referred to as the female part of a perfect flower.

■ plant cell culture
growth of plant cells or roots of plants in bioreactors.

■ plant cell immobilisation
entrapment of plant cells in gel matrices; the cells are suspended in small drops of the material, which then is set or allowed to harden to make little carriers. Materials such as alginates, agar or polyacrylamide can be used.

■ plantlet
a small rooted shoot developed from seed or from cultured cells either by embryogenesis or organogenesis.

■ plaque
a clear spot on an otherwise opaque culture plate of bacteria or cultured bacteria cells, showing where cells have been lysed by viral infection.

■ plasma cells
antibody-producing white blood cells derived from B lymphocytes.

■ plasma membrane
see **cell membrane; plasmalemma.**

■ plasmalemma
a delicate cytoplasmic double membrane found on the outside of the protoplast, adjacent to the cell wall. *See* cell membrane.

■ plasmid
an extrachromosomal, autonomous circular DNA molecule found in certain bacteria, capable of autonomous replication. Plasmids can transfer genes between bacteria and are important tools of transformation in genetic engineering. They exist in an autonomous state and are transferred independently of chromosomes. *See* nicked circle; pAMP; relaxed plasmid; stringent plasmid; supercoiled plasmid.

■ plastid
a cytoplasmic body found in the cells of plants and some protozoa. Chloroplastids, for example, produce chlorophyll that is involved in photosynthesis.

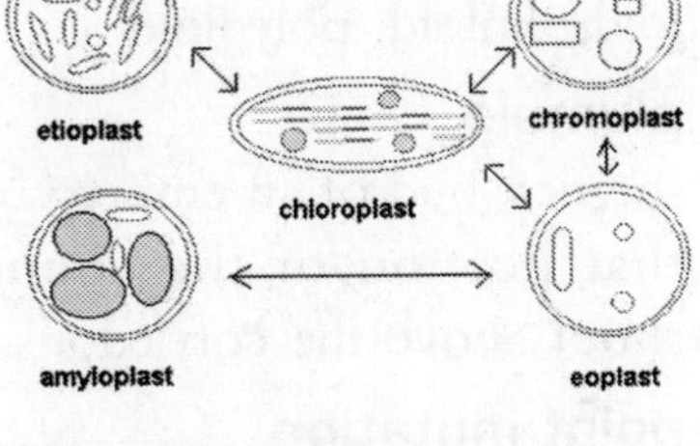

■ plastoquinone
a quinone which is one of a group of compounds involved

in the transport of electrons in photosynthesis in chloroplasts.

■ **plate**

1. verb: to distribute a thin film of something. Hence micro-organisms or plant cells are plated onto nutrient agar. 2. noun: refers to the two segments of a Petri dish or a similar-shaped item.

■ **plating efficiency**

the percentage of inoculated cells which give rise to cell colonies when seeded into culture vessels.

■ **pleiotropy**

the situation in which a particular gene has an effect on several different traits.

■ **ploidy**

the number of sets of chromosomes per cell, e.g., haploid, diploid, polyploid.

■ **plumule**

the first bud of an embryo or that portion of the young shoot above the cotyledons.

■ **point mutation**

a change in DNA at a specific site in a chromosome. Includes nucleotide substitutions and the insertion or deletion of one or a few nucleotide pairs. *See* mutation.

■ **polar bodies**

in female animals, the products of a meiotic division that do not develop into an ovum. The first polar body comprises one of the two products of meiosis I, and it may not go through meiosis II. The second polar body comprises one of the two products of meiosis II.

■ **polar mutation**

a mutation that influences the functioning of genes that are downstream in the same transcription unit.

■ **polar nuclei**

two centrally located nuclei in the embryo sac that unite with a second sperm cell in a triple fusion. In certain seeds, the product of this fusion develops into the endosperm.

■ **polar transport**

the directed movement within plants of compounds (usually endogenous plant growth regulators) mostly in one direction; polar transport overcomes the tendency for diffusion in all directions.

■ **polarity**
the observed differentiation of an organism, tissue or cell into parts having opposed or contrasted properties or form.

■ **pole cells**
a group of cells in the posterior of *Drosophila* embryos that are precursors to the adult germ line.

■ **pollen**
the mass of germinated microspores or partially developed male gametophytes of seed plants.

■ **pollen culture**
the *in vitro* culture and germination of pollen grains. Callus cultures thus obtained will form shoots or embryoids which develop into monoploid plants. *See* anther culture; microspore culture.

■ **pollen grain**
a microspore produced in the pollen sac of angiosperms or the microsporangium of gymnosperms. Unicellular, with variable shape and size, and usually ovoid from 25 to 250mm.

■ **pollination**
transfer of pollen from anther to stigma in the process of fertilisation in angiosperms; transfer of pollen from male to female cone in the process of fertilisation in gymnosperms.

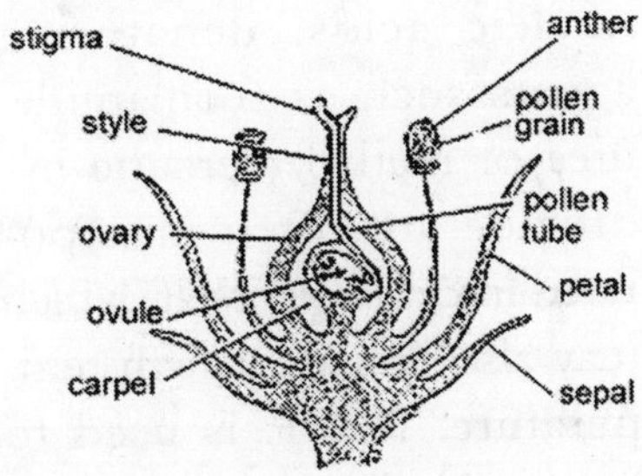

■ **poly (A) polymerase**
enzyme that catalyses the addition of adenine residues to the 3´ end of pre-mRNAs to form the poly-(A) tail. *See* polyadenylation; polymerase.

■ **polyacrylamide gel electrophoresis**
a method for separating nucleic acid or protein molecules according to their molecular size. The molecules migrate through the inert gel matrix under the influence of an electric field. In the case of protein PAGE, detergents such as sodium dodecyl sul-

phate (SDS) are often added to ensure that all molecules have a uniform charge. Secondary structure can often lead to the anomalous migration of molecules. Therefore it is common to denature protein samples by boiling them prior to PAGE. In the case of nucleic acids, denaturing agents such as formamide, urea or methyl mercuric hydroxide are often incorporated into the gel itself, which may also be run at high temperature. PAGE is used to separate the products of DNA-sequencing reactions and the gels employed are highly denaturing, since molecules differing in size by a single nucleotide must be resolved.

■ polyacrylamide gels

often referred to incorrectly as acrylamide gels. These gels are made by cross-linking acrylamide with *N,N'*-methylene-*bis*-acrylamide. Polyacrylamide gels are used for the electrophoretic separation of proteins, DNA and RNA molecules. Polyacrylamide beads are also used as molecular sieves in gel chromatography, marketed as Bio-gelTM.

■ polyadenylation

post-transcriptional addition of a polyadenylic acid tail to the 3′ end of eukaryotic mRNAs. Also called poly-(A) tailing. The adenine-rich 3′ terminal segments is called a poly (A) tail.

■ polyclonal antibody

a serum sample that contains a mixture of immunoglobulin molecules secreted against a specific antigen, each recognising a different epitope, some antibodies of which bind to different antigenic determinants of one antigen.

■ polyembryony

in the ordinary course of events, one embryo is formed in each ovule and is derived from the fertilisation of the ovum in the solitary embryo sac. However, two or more embryos could start development even though only one may reach maturity.

■ polyethylene glycol

a polymer having the general formula: $HOCH_2(CH_2OCH_2)XCH_2OH$ and available in a range of molecular weights from *ca* 1000 to *ca* 6000. PEG 4000 and PEG 6000 are commonly used to promote cell or protoplast fusion, and to facilitate DNA uptake in the transformation of organisms such as yeast. PEG is also used to concentrate solutions by withdrawing water from them.

■ polygene

one of many genes of small effect influencing the development of a quantitative trait; results in continuous variation and in quantitative inheritance. *See* gene.

■ polygenic

controlled by many genes of small effect.

■ polylinker

a segment of DNA that contains a number of different restriction endonuclease sites. a.k.a. multiple cloning site (MCS).

■ polymer

a compound composed of many identical smaller subunits; resulting from the process of polymerization.

■ polymerase

an enzyme that catalyses the formation of polymeric molecules from monomers. A DNA polymerase synthesises DNA from deoxynucleoside triphosphates using a complementary DNA strand and a primer. An RNA polymerase synthesises RNA from monoribonucleoside triphosphates and a complementary DNA strand. *See* poly-(A) polymerase; polymerase chain reaction; RNA polymerase; *taq* polymerase.

■ polymerase chain reaction

a procedure that amplifies a particular DNA sequence. It involves multiple cycles of *denaturation*, *annealing* to oligonucleotide primers, and *extension* (polynucleotide synthesis), using a thermostable DNA polymerase, deoxyribonucleotides, and primer sequences in multiple cycles of denaturation-renaturation-DNA synthesis. *See* polymerase.

■ **polymerisation**
chemical union of two or more molecules of the same kind such as glucose or nucleotides to form a new compound (starch or nucleic acid) having the same elements in the same proportions but a higher molecular weight and different physical properties.

■ **polymery**
the phenomenon whereby a number of genes at different loci (which may be polygenes) can act together to produce a single effect.

■ **polymorphism**
the occurrence of two or more alleles at a locus in a population. a.k.a. genetic polymorphism.

■ **polynucleotide**
a chain of nucleotides in which each nucleotide is linked by a single phospho-diester bond to the next nucleotide in the chain. They can be double- or single-stranded. The term is used to describe DNA or RNA. *See* nucleotide.

■ **polypeptide**
a linear molecule composed of two or more amino acids linked by covalent (peptide) bonds. They are called dipeptides, tripeptides and so forth, according to the number of amino acids present.

■ **polyploid**
tissue or cells with more than two complete sets of chromosomes, that results from chromosome replication without nuclear division or from union of gametes with different number of chromosome sets, hence triploid (3*x*), tetraploid (4*x*), pentaploid (5*x*), hexaploid (6*x*), heptaploid (7*x*), octoploid (8*x*)).

■ **polysaccharide**
long-chain molecules, such as starch and cellulose, composed of multiple units of a monosaccharide.

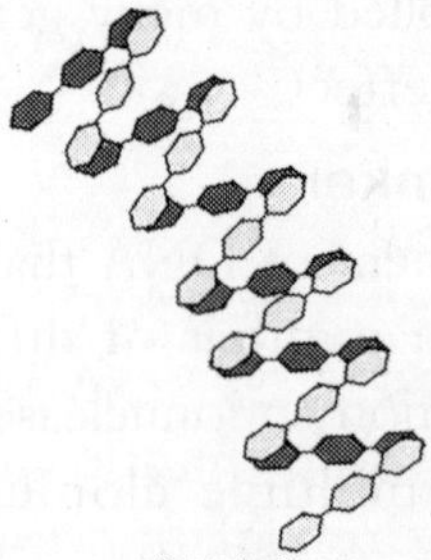

polysaccaharide

■ **polysaccharide capsules**
carbohydrate coverings with antigenic specificity that are

present on some types of bacteria.

polyspermy

the entry of several sperm into the egg during fertilisation, although only one sperm nucleus actually fuses with the egg nucleus.

polytene chromosomes

giant chromosomes produced by interphase replication without division, and consisting of many identical chromatids arranged side by side in a cablelike pattern.

polyunsaturates

oils containing polyunsaturated fatty acids.

polyvalent vaccine

a recombinant organism into which antigenic determinants have been cloned from a number of different disease-causing organisms, and used as a vaccine. *See* vaccine.

polyvinylpyrrolidone

(PVP) C_6H_9NOnAn occasional constituent of plant tissue culture isolation media. PVP is of variable molecular weight and has antioxidant properties, so is used to prevent oxidative browning of explanted tissues. PVP is less frequently used as an osmoticum in culture media.

population

a local group of organisms belonging to the same species and interbreeding.

population density

number of cells or individuals per unit. The unit could be an area or volume of medium.

population genetics

the branch of genetics that deals with frequencies of alleles and genotypes in breeding populations. *cf.* quantitative genetics.

Porcine Endogenous Retrovirus (PERV)

the provirus (q.v.) of a porcine retrovirus. With the increasing interest in the use of pig organs for xenotransplantation to humans, there has been concern that PERVs could be activated after transplantation, creating an infection in the human recipient.

position effect

the situation in which a change in phenotype results from the change of the position of a gene or group of genes.

■ **positional candidate gene**
a gene known to be located in the same region as a DNA marker that has been shown to be linked to a single-locus trait or to a quantitative trait locus (QTL) (q.v.) and whose function suggests that it could be the source of genetic variation in the trait in question.

■ **positional cloning**
the process that commences with searching for markers linked to a particular inherited trait, then using those markers to identify the approximate location of the gene responsible for the trait, and then using various cloning strategies (including chromosome walking, q.v.) to identify, isolate and characterise the gene. Originally, the strategy was called *reverse genetics*, a term that some consider to be a misnomer and misleading.

■ **positive control system**
a mechanism in which the regulatory protein(s) is required to turn on gene expression.

■ **primary immune response**
the immune response that occurs during the first encounter of a mammal with a given antigen. *cf* secondary immune response.

■ **primary meristem**
meristem of the shoot or root tip giving rise to the primary plant body.

■ **primary tissue**
a tissue that has differentiated from a primary meristem.

■ **primary transcript**
the RNA molecule produced by transcription prior to any post-transcriptional modifications; also called a pre-mRNA in eukaryotes.

■ **primer**
a short DNA or RNA fragment annealed to a template of single-stranded DNA, providing a 3′ hydroxyl end from which DNA polymerase extends a new DNA strand to produce a duplex molecule.

■ **primer DNA polymerase**

DNA polymerase that provides primers for the DNA polymerization. Unlike RNA polymerase, DNA polymerase is unable to initiate the *de novo* synthesis of a polynucleotide chain. DNA polymerase can only add nucleotides to a free 3´ hydroxyl group at the end of a pre-existing chain. A short oligonucleotide, known as a primer, is therefore needed to supply such a hydroxyl group for the initiation of DNA synthesis.

■ **primer walking**

a method for sequencing long (>1 kb) cloned pieces of DNA. The initial sequencing reaction reveals the sequence of the first few hundred nucleotides of the cloned DNA. On the basis of these data, a primer containing about 20 nucleotides and complementary to a sequence near the end of sequenced DNA is synthesised, and is then used for sequencing the next few hundred nucleotides of the cloned DNA. This procedure is repeated until the complete nucleotide sequence of the cloned DNA is determined.

■ **probe DNA**

a labelled DNA molecule used to detect complementary-sequence nucleic acid molecules by molecular hybridisation. To localize the probe DNA sequence and reveal the complementary hybridisation sequence, autoradiography or fluorescence is used. *See* multilocus probe; nucleotide.

■ **procambium**

a primary meristem that gives rise to primary vascular tissues and, in most woody plants, to the vascular cambium.

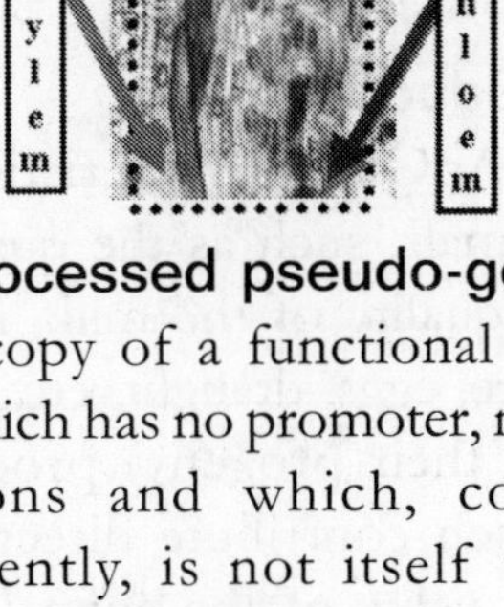

■ **processed pseudo-gene**

a copy of a functional gene which has no promoter, no introns and which, consequently, is not itself transcribed. Pseudo-genes are thought to originate from the integration into the genome of cDNA copies synthesised from mRNA molecules by reverse transcriptase. Pseudo-genes therefore have a poly (dA) sequence at their 5´ ends. Because they are not subject to any evolutionary pressure to maintain their coding potential, pseudo-genes accumulate mutations and often have stop codons in all three reading frames.

■ **production environment**

in AnGR: all input-output relationships, over time, at a particular location. The relationships will include biological, climatic, economic, social, cultural and political factors, which combine to determine the productive potential of a particular livestock enterprise.

■ **production traits**

in AnGR: characteristics of animals, such as the quantity or quality of the milk, meat, fibre, eggs, draught, etc., they (or their progeny) produce, which contribute directly to the value of the animals for the farmer, and that are identifiable or measurable at the individual level. Production traits of farm animals ate generally quantitatively inherited, i.e., they are influenced by many genes whose expression in a particular animal also reflects environmental influences. (Source: FAO, 1999).

■ **productivity**

the amount of product that is produced within a given period of time from a specified quantity of resource.

■ **pro-embryo**

a group of cells arising from the division of the fertilised egg cell or somatic embryo before those cells which are to become the embryo are recognisable.

■ **progeny testing**

for a single locus: the practice of ascertaining the genotype of an individual from its offspring, such as by mating it to other individuals and examining the progeny. For a quantitative trait: the collection and use of progeny performance as a clue to the breeding value of the parent of those progeny.

■ **progesterone**

a hormone produced primarily by the corpus luteum of the ovary, but also by the placenta, that prepares the inner lining of the uterus for implantation of a fertilised egg cell.

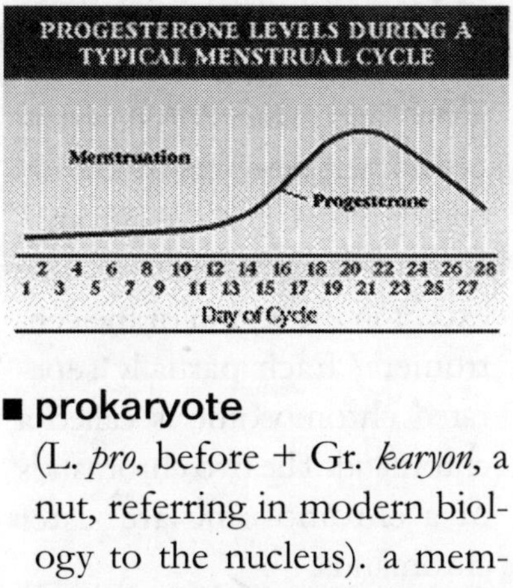

■ **prokaryote**

(L. *pro*, before + Gr. *karyon*, a nut, referring in modern biology to the nucleus). a member of a large group of organisms, including bacteria and blue-green algae, which do not have the DNA separated from the cytoplasm by a membrane in their cells. The DNA is usually in one long strand. Prokaryotes do not undergo meiosis and do not have functional organelles such as mitochondria and chloroplasts. *cf* eukaryote.

■ **prolactin**

a hormone, produced by the anterior pituitary gland, that stimulates and controls lactation in mammals.

■ **proliferation**

increase by frequent and repeated reproduction; growth by cell division.

■ **pro-meristem**

the embryonic meristem that is the source of organ initials or foundation cells.

■ **pro-nuclear micro-injection**

the initial method of transgenesis in animals, involving injecting many copies of a particular gene into one of the two pro-nuclei of a fertilised ova. Characterised by a very low success rate. As animal cloning (q.v.) becomes more common, pro-nuclear micro-injection will be replaced by micro-injection of cloned genes into a culture of cloned embryos produced by nuclear transfer, which can be tested for expression of the transgene before transfer to recipient females. See **transgenesis.**

■ pro-nucleus

either of the two haploid gamete nuclei just prior to their fusion in the fertilised ovum.

■ proofreading

the enzymatic scanning of newly-synthesised DNA for structural defects, such as mis-matched base pairs. It is effected by DNA polymerase.

■ propagation

the multiplication of plants by numerous types of vegetative material; an ancient practice dating from the dawn of agriculture, carried out in a nursery or directly in the field (vegetative propagation), and now in *in vitro* culture (micropropagation).

■ propagule

any structure capable of giving rise to a new plant by asexual or sexual reproduction, including bulbils, leafbuds, etc.

■ pro-phage

the genome of a temperate bacteriophage integrated into the chromosome of a lysogenic bacterium and replicated along with the host chromosome.

■ prophase

an early stage in nuclear division, characterised by the shortening and thickening of the chromosomes and their movement to the metaphase plate. It occurs between interphase and metaphase. During this phase, the centriole divides and the two daughter centrioles move apart. Each sister DNA strand from interphase replication becomes coiled, and the chromosome is longitudinally double except in the region of the centromere. Each partially separated chromosome is called a chromatid. The two chromatids of a chromosome are sister chromatids.

Prophase

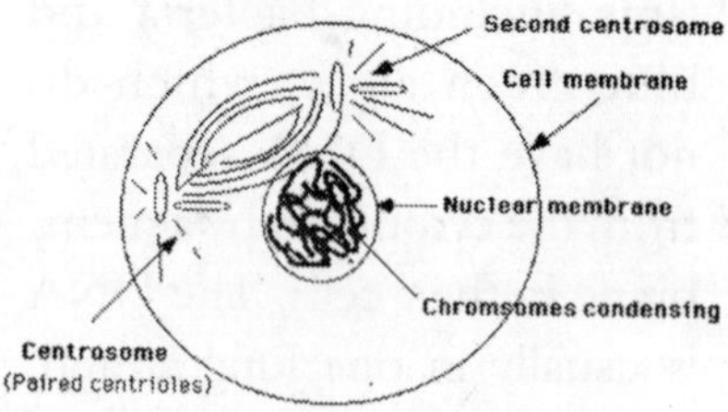

■ protamines

small basic proteins that replace the histones in the chromosomes of some sperm cells.

■ protease

an enzyme that hydrolyzes proteins, cleaving the peptide bonds that link amino acids in protein molecules.

■ protocorm

in orchids, seeds contain an unorganised embryo comprising only a few hundred cells. During seed germination this embryo forms a tuberous structure, called a protocorm, from which develops a complete plant. In cultures, vegetative parts of several orchids form round, smooth protocorms which may be multiplied indefinitely or induced to regenerate a whole plant.

■ protoderm

a primary meristem tissue that gives rise to epidermis.

■ protogyny

the condition in which the female reproductive organs (carpels) of a flower mature before the male ones (stamens), thereby ensuring that self-fertilisation does not occur.

■ proto-oncogene

a normal cellular gene that can be changed to an oncogene by mutation or by being incorporated into a retrovirus and then being transcribed at inappropriate times and/or in inappropriate tissues.

■ protoplasm

the essential, complex living substance of cells, upon which all vital functions of nutrition, secretion, growth and reproduction depend.

■ protoplast

a bacterial or plant cell for which the relatively rigid wall has been removed either chemically or enzymatically, leaving its cytoplasm enveloped by only a delicate peripheral membrane. Protoplasts are spherical and smaller than the elongate, angular shaped and often vacuolated cells from which they have been released.

■ protoplast culture

the isolation and culture of plant protoplasts by mechanical means or by enzymatic digestion of plant tissues or organs, or cultures derived from these. Protoplasts are utilized for selection or hybridisation at the cellular level and for a variety of other purposes.

■ protoplast fusion

the coalescence of the plasmalemma and cytoplasm of two or more protoplasts in contact with one another. Initial adhesion of two protoplasts is a random process but coalescence can be induced.

■ prototroph

an organism such as a bacterium that will grow on a minimal medium.

■ pro-toxin

a latent, non-active precursor form of a toxin.

■ protozoan

a microscopic, single-cell organism.

■ provirus

a retrovirus in which the single RNA strand has been converted into a double-strand DNA, which has been integrated into a host genome.

■ pseudo-autosomal region

a section at one end of the X and Y chromosomes for which there is sufficient homology that there is pairing (synapsis) during meiosis I.

■ quantitative genetics

the area of genetics concerned with the inheritance of continuously-varying traits. Most practical improvement programs involve the application of quantitative genetics.

■ quantitative inheritance

inheritance of measurable traits (height, weight, colour intensity, etc.) that depend on the cumulative action of many genes.

■ quantitative trait

a measurable trait that shows continuous variation; a trait that can not be classified into a few discrete classes.

■ quantitative trait locus

a locus that affects a quantitative trait. The plural form (quantitative trait loci) is also abbreviated as QTL.

■ quantum

an elemental unit of energy. Its energy value is hv, where

h, Planck's constant, is 6.62×10^{-27} erg-second and v is the frequency of the vibrations or waves with which the energy is associated. *cf* photon.

quantum speciation

the rapid formation of new species, primarily by genetic drift (q.v.).

quarantine

originally, keeping a person or living organism in isolation for a period (originally 40 days) after arrival to allow disease symptoms to appear, if there was any disease present. Now used for regulations restricting the sale or shipment of living organisms, usually to prevent disease or pest invasion of an area.

quiescent

quiet, at rest, but not necessarily dormant, and having the potential for resumed activity; can apply to non-meristematic cells.

race

a distinguishable group of organisms of a particular species, that are geographically, ecologically, physiologically, physically and/or chromosomically distinct from other members of the species.

■ raceme

an inflorescence in which the main axis is elongated but the flowers are borne on pedicels that are about equal in length.

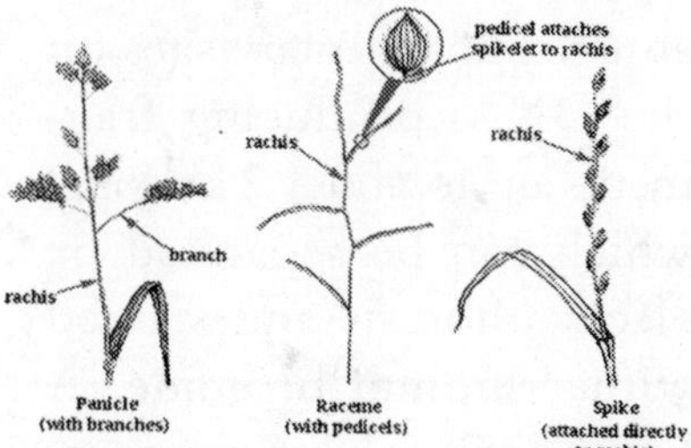

■ rachis

main axis of a spike; axis of fern leaf (frond) from which pinnae arise; in compound leaves, the extension of the petiole corresponding to the midrib of an entire leaf.

■ radicle

that portion of the plant embryo which develops into the primary or seed root.

■ radioactive isotope, radioisotope

an unstable isotope that emits ionising radiation.

ramet
an individual member of a clone. *cf* ortet.

random amplified polymorphic DNA
a technique using single, short (usually 10-mer) synthetic oligonucleotide primers for PCR. The primer, whose sequence has been chosen at random, initiates replication at its complementary sites on the DNA, producing fragments up to about 2 kb long, which can be separated by electrophoresis and stained with ethidium bromide. A primer can exhibit polymorphism between individuals, and polymorphic fragments can be used as markers.

random mutagenesis
a non-directed change of one or more nucleotide pairs in a DNA molecule.

random primer method
a protocol for labelling DNA *in vitro*. A sample of random oligonucleotides (6 or 14 nucleotides long) containing all possible combinations of nucleotide sequences is hybridised to a DNA probe. Then, in the presence of a DNA polymerase and the four deoxyribonucleotides - one of which is labelled - the 3′ hydroxy ends of the hybridised oligonucleotides provide initiation sites for DNA synthesis that uses the separated strands of the probe DNA as a template. This reaction produces labelled copies of portions of the probe DNA.

reading frame
a series of triplets beginning from a specific nucleotide. Each triplet is represented by a single amino acid in the protein synthesised. The reading frame defines which sets of three nucleotides are read as triplets in the DNA, and hence as codons in the corresponding mRNA; this is determined by the initiation codon, AUG. Thus the sequence
AUGGCAAAAUUUCCC
would read as
AUG/GCA/AAA/UUU/CCC/ and not as
A/UGC/CAA/AAU/UUC/CC.
depending on where one begins, each DNA strand con-

tains three different reading frames. *See* open reading frame, overlapping reading frames.

■ read-through
transcription or translation that proceeds beyond the normal stopping point because of the absence of the transcription or translation termination signal of a gene.

■ reca
a protein in most bacteria, and that is essential for DNA repair and DNA recombination.

■ recalcitrant
of seeds: unable to survive drying and subsequent storage at low temperature. *See* field gene bank.

■ receptacle
enlarged end of the pedicel or peduncle, to which other flower parts are attached.

■ receptor
a molecule that can accept the binding of a ligand.

■ receptor-binding screening
one of the biotechnology-based methods for discovering conventional drugs. The method relies on the fact that many drugs act by binding to specific proteins (receptors) on or in cells: these proteins usually bind to hormones or to other cells, and control the cell's behaviour, although they may be enzymes or structural elements of the cell. The drug interferes with the normal role of the protein.

■ recessive
describing an allele whose effect with respect to a particular trait is not evident in heterozygotes. Opposite to *dominant*.

■ recessive oncogene
a single copy of this gene is sufficient to suppress cell proliferation; the loss of both copies of the gene contributes to cancer formation. *See* oncogene.

■ reciprocal crosses
crosses between the same two strains, but with the sexes reversed; e.g., female A × male B and male A × female B.

■ reciprocating shaker
a platform shaker used for agitating culture flasks, with

a back and forth action at variable speeds.

■ recognition site

a nucleotide sequence - composed typically of 4, 6 or 8 nucleotides - that is recognised by and to which a restriction endonuclease (restriction enzyme) binds. For type II restriction enzymes (those used in gene-cloning experiments) it is also the sequence within which the enzyme specifically cuts (and their corresponding enzymes methylate) the DNA, i.e., for type II enzymes, the recognition site and the target site are the same sequence. Type I enzymes bind to their recognition site and then cleave the DNA at some more or less random position outside that recognition site. *cf* restriction site.

■ recombinant

a term used in both classical and molecular genetics.

1. in classical genetics: an organism or cell that is the result of recombination (crossing-over), e.g., Parents: *AB*/*ab* and *ab*/*ab*; recombinant offspring: *Ab*/*ab*.

2. In molecular genetics: A molecule containing DNA from different sources. The word is typically used as an adjective, e.g., recombinant DNA.

■ recombinant DNA

the result of combining DNA fragments from different sources.

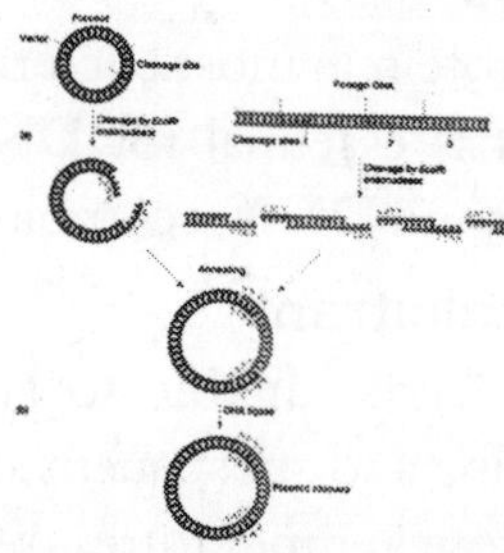

■ recombinant DNA technology

a set of techniques which enable one to manipulate DNA. One of the main techniques is DNA cloning (because it produces an unlimited number of copies of a particular DNA segment), and the result is sometimes called a DNA clone or gene clone (if the segment is a gene), or simply a clone. An organism manipulated using recombinant DNA techniques is called a genetically modified organism (GMO).

among other things, recombinant DNA technology involves:

- identifying genes;
- cloning genes;
- studying the expression of cloned genes; and
- producing large quantities of the gene product.

■ recombinant protein

a protein whose amino acid sequence is encoded by a cloned gene.

■ recombinant RNA

a term used to describe RNA molecules joined *in vitro* by T4 RNA ligase.

■ recombinant toxin

a single multifunctional toxic protein that has been created by combining the coding regions of various genes.

■ recombinant vaccine

a vaccine produced from a cloned gene.

■ recombination

the process of crossing over, which occurs during meiosis I. It involves breakage in the same position of each of a pair of non-sister chromatids from homologous chromosomes, followed by joining of non-sister fragments, resulting in a reciprocal exchange of DNA between non-sister chromatids within an homologous pair of chromosomes.

■ reconstructed cell

a viable transformed cell resulting from genetic engineering.

■ re-differentiation

cell or tissue reversal from one differentiated type to another differentiated type of cell or tissue. *See* de-differentiation.

■ reduction division

phase of meiosis in which the maternal and paternal chromosomes of the bivalent separate.

■ regeneration

the growth of new tissues or organs to replace those injured or lost. In tissue culture, regeneration is used to define the development of organs or plantlets from a tissue, callus culture or from a bud. *See* conversion; micropropagation; organogenesis.

■ regulator
substance regulating growth and development of cells, organs, etc.

■ regulatory gene
a gene whose protein controls the activity of other genes or metabolic pathways.

■ rejuvenation
reversion from adult to juvenile stage.

■ relaxed plasmid
a plasmid that replicates independently of the main bacterial chromosome and is present in 10-500 copies per cell. *See* plasmid.

■ release factors
1. soluble protein that recognises termination codons in mRNAs and terminate translation in response to these codons.
2. a hormone that is produced by the hypothalamus and stimulates the release of a hormone from the anterior pituitary gland into the bloodstream.

■ re-naturation
the re-association of two nucleic acid strands after denaturation. The restoration of a molecule to its native form. In nucleic acid biochemistry, this term usually refers to the formation of a double-stranded helix from compleme tary single-strandd molecules. Some simple proteins can also be re-natured and regain their function.

■ rennin
an enzyme secreted by cells lining the stomach in mammals, and that is responsible for clotting milk.

■ repeat unit
a sequence of bases that occurs repeatedly in the genome, often end-on-end, i.e., tandemly.

■ repetitive DNA
DNA sequences that are present in a genome in multiple copies, sometimes a million times or more.

■ replacement therapy
the administration of metabolites, co-factors or hormones that are deficient as the result of a genetic disease.

■ **replacement, gene replacement**

a method of substituting a cloned gene, or part of a gene, which may have been mutated *in vitro*, for the wild-type copy of the gene within the host's chromosome. *See* homogenotization.

■ **replica plating**

a procedure for duplicating the bacterial colonies growing on agar medium in one Petri plate to agar medium in another Petri plate.

■ **replication**

the synthesis of duplex (double-stranded) DNA by copying from a single-stranded template.

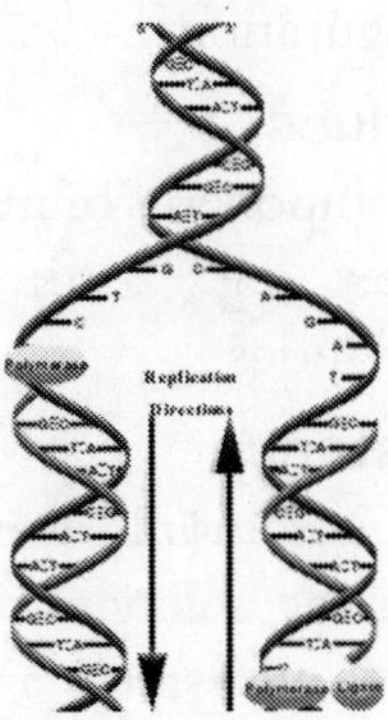

■ **replicative form**

the molecular configuration of viral nucleic acid that is the template for replication in the host cell.

■ **replicon**

the portion of a DNA molecule which is replicable from a single origin. Plasmids and the chromosomes of bacteria, phages and other viruses usually have a single origin of replication and, in these cases, the entire DNA molecule constitutes a single replicon. Eukaryotic chromosomes have multiple internal origins and thus contain several replicons. The word is often used in the sense of a DNA molecule capable of independent replication, e.g., "The shuttle vector pJDB219 is a replicon in both yeast and *E. coli*."

■ **replisome**

the complete replication apparatus present at a replication fork that carries out the semi-conservative replication of DNA.

■ **reporter gene**

a gene that encodes a product that can readily be assayed. Thus reporter genes are used to determinate whether a particular DNA construct has been successfully introduced into a cell, organ or tissue.

repressible enzyme
an enzyme whose synthesis is diminished by a regulatory molecule.

repression
inhibition of transcription by preventing RNA polymerase from binding to the transcription initiation site: a repressed gene is "turned off."

repressor
a protein which binds to a specific DNA sequence (the operator) upstream from the transcription initiation site of a gene or operon and prevents RNA polymerase from commencing mRNA synthesis. Examples of repressors are the C_1 protein of bacteriophage and the *lac*1 protein of the *lac* operon.

reproduction
the production of an organism, cell or organelle like itself (self propagation).
1. Sexual reproduction: the regular alternation (in the lifecycle of haplontic, diplontic and diplohaplontic organisms) of meiosis and fertilisation (karyogamy) which provides for the production of offspring. The main biological significance of sexual reproduction lies in the fact that it achieves genetic recombination.
2. Asexual or agamic reproduction: the development of a new individual from either a single cell (agamospermy) or from a group of cells (vegetative reproduction) in the absence of any sexual process.

repulsion
the phase state in which a dominant (or wild-type) allele at one locus and a recessive (or mutant) allele at a second locus occur on the same chromosome. Also called trans configuration.

residues
the components of macromolecules, e.g., amino acids, nucleotides.

resistance
term commonly used to describe the ability of an organism to withstand a stress, a force or an effect of a disease, or its agent or a toxic substance.

■ resistance factor

a plasmid that confers antibiotic resistance to a bacterium.

■ rest period

an endogenous physiological condition of viable seeds, buds or bulbs that prevents growth even in the presence of otherwise favourable environmental conditions. By some seed physiologists, this is referred to as dormancy.

■ restitution nucleus

a nucleus with unreduced or doubled chromosome number that results from the failure of a meiotic or mitotic division.

■ restriction endonuclease

a class of endonucleases that cleaves DNA after recognising a specific sequence, e.g., *Bam*H1 (5´GGATCC3´), *Eco*RI (5´GAATTC3´), and *Hin*dIII (5´AAGCTT3´). There are three types of restriction endonucleas enzymes:

Type I: Cuts non-specifically a distance greater than 1000 bp from its recogniti n sequence and contains both restriction and methylation activities.

Type II: Cuts at or near a short, and often palindromic (q.v.) , recognition sequence. A separate enzyme methylates the same recognition sequence. They may make the cuts in the two DNA strands exactly opposite one another and generate blunt ends, or they may make staggered cuts to generate sticky ends. The type II restriction enzymes are the ones commonly exploited in recombinant DNA technology.

Type III: Cuts 24-26 bp downstream from a short, asymmetrical recognition sequence. Requires ATP and contains both restriction and methylation activities.

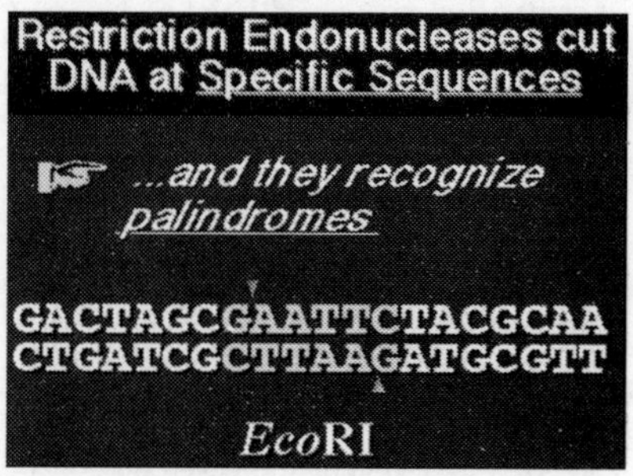

■ restriction exonuclease

a class of nucleases that degrades DNA or RNA, starting from an end either 5´ or 3´.

■ **restriction fragment**
a fragment of DNA produced by cleaving (digesting, cut ing) a DNA molecule with one or more restriction endonucleases.

■ **restriction fragment length polymorphism**
the occurrence of variation in the length of DNA fragments that are produced after cleavage with a type II restriction endonuclease. The differences in DNA lengths are due to the presence or absence of recognition site(s) for that particular restriction enzyme. RFLPs were initially detected using hybridisation with DNA probes after separation of digested genomic DNA by gel electrophoresis (Southern analysis). Now they are typically detected by electrophoresis of digested PCR product.

■ **restriction map**
the linear array of restriction endonuclease sites on a DNA molecule.

■ **restriction nuclease**
a bacterial enzyme that cuts DNA at a specific site.

■ **reticulocyte**
a young red blood cell.

■ **retro-element**
any of the integrated retroviruses or the transposable elements that resemble them.

■ **retro-poson**
a transposable element that moves via reverse transcription (i.e., from DNA to RNA to DNA) but lacks the long terminal repeat sequences.

■ **retroviral vectors**
gene transfer systems based on viruses that have RNA as their genetic material.

■ **retrovirus**
a class of eukaryotic RNA viruses that can form double-stranded DNA copies of their genomes by using reverse transcription; the double-stranded forms integrate into chromosomes of an infected cell. Many naturally occurring cancers of vertebrate animals are caused by retroviruses. Also, the AIDS virus is a retrovirus.

■ **reversal transfer**
transfer of a culture from a callus-supporting medium to a shoot-inducing medium.

■ **reverse mutation**
restitution of a mutant gene to the wild-type condition, or at least to a form that gives the wild phenotype; more generally, the appearance of a trait expressed by a remote ancestor.

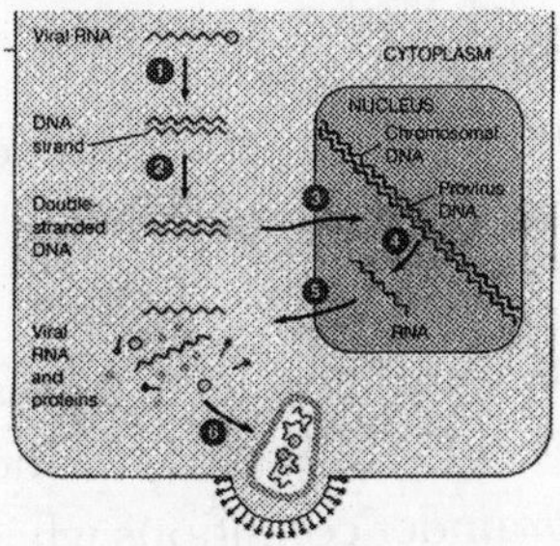

■ **reverse transcriptase**
an enzyme that uses RNA molecule as a template for the synthesis of a complementary DNA strand.

■ **reverse transcription**
the synthesis of DNA on a template of RNA, accomplished by reverse transcriptase.

■ **rhizobacterium**
a micro-organism whose natural habitat is near, on or in plant roots.

■ **rhizobium**
prokaryote able to establish symbiotic relationship with leguminous plants, as a result of which elemental nitrogen is fixed or converted to ammonia. *See* nitrogen fixation.

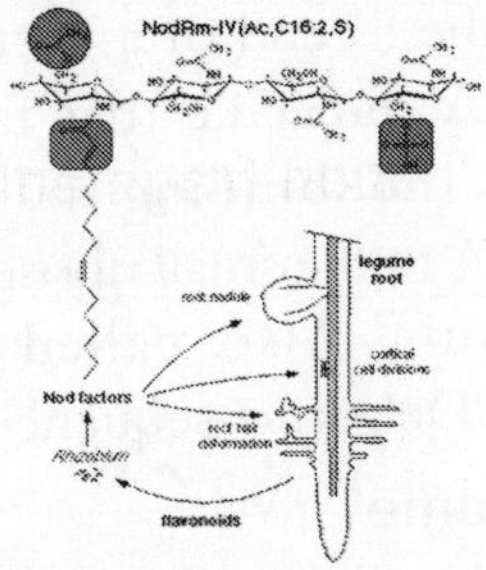

■ **rhizosphere**
the soil region in the immediate vicinity of growing plant roots.

■ **Ri plasmid**
a class of large conjugative plasmids found in the soil bacterium *Agrobacterium rhizogenes*. Ri plasmids are responsible for hairy root disease of certain plants. A segment of the Ri plasmid is found in the genome of tumour tissue from plants with hairy root disease.

■ **ribonuclease**
any enzyme that hydrolyses RNA.

■ **ribose**
a monosaccharide ($C_5H_{10}O_5$) rarely occurring free in nature,

but important as a component of RNA.

■ **ribosomal binding site**

a sequence of nucleotides near the 5′ end of a bacterial mRNA molecule that facilitates the binding of the mRNA to the small ribosomal sub-unit. Also called the Shine-Delgarno sequence.

■ **ribosome**

the sub-cellular structure that contains both RNA and protein molecules and mediates the translation of mRNA into protein. Ribosomes comprise large and small sub-units. *See* organelle; translation.

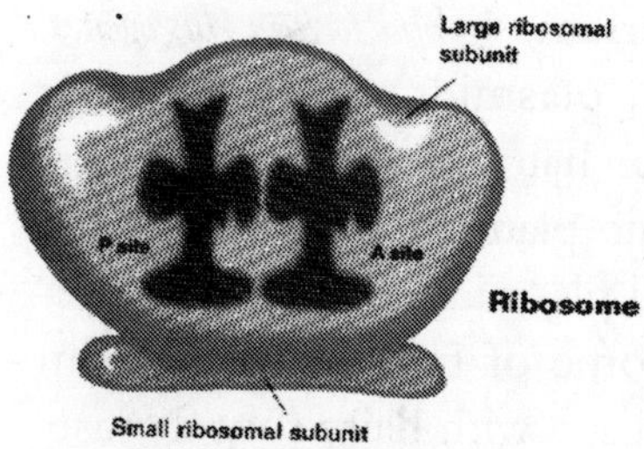

■ **ribozyme**

RNA molecule that can catalyse chemical reactions, often cutting other RNAs.

■ **ribulose**

a keto-pentose sugar ($C_5H_{11}O_5$) that is involved in carbon dioxide fixation in photosynthesis.

■ **ribulose biphosphate**

a five-carbon sugar that is combined with carbon dioxide to form a six-carbon intermediate in the first stage of the dark reaction of photosynthesis.

■ **rinderpest**

cattle plague; a viral infection of cattle, sheep and goats.

■ **R-loops**

single-stranded DNA regions in RNA-DNA hybrids formed *in vitro* under conditions where RNA-DNA duplexes are more stable than DNA-DNA duplexes.

■ **RNA**

Ribonucleic acid. an organic acid composed of repeating nucleotide units of adenine, guanine, cytosine and uracil, whose ribose components are linked by phospho-diester bonds. The information-carrying material in some viruses. More generally, a molecule derived from DNA by transcription that may carry information (messenger RNA (mRNA)), provide sub-cellular structure (ribosomal RNA (rRNA)), transport amino acids (transfer RNA (tRNA)) or facilitate the biochemical modification of itself or other RNA molecules. *See* antigen RNA; gene splicing; heterogeneous nuclear RNA (hnRNA); mRNA; ribosomal RNA; RNA polymerase; small nuclear RNA; transfer RNA.

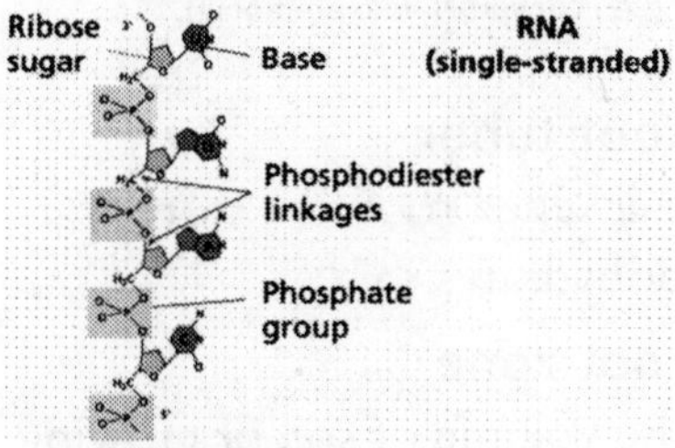

■ RNA editing

post-trancriptional processes that alter the information encoded in gene transcripts (RNAs).

■ RNA polymerase

an enzyme that catalyses the synthesis of RNA from a DNA template. *See* polymerase; RNA.

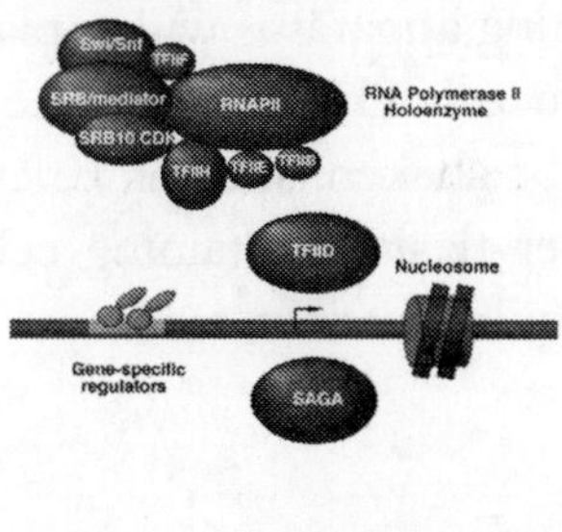

■ RNase

Ribonuclease. a group of enzymes that catalyse the cleavage of nucleotides in RNA.

■ Roentgen

(symbol: r) obsolete unit of ionising radiation. The SI unit is the sievert (symbol: Sv; 1 Sv » 8.4 r).

■ root

the descending axis of a plant, normally below ground, which serves to anchor the plant and to absorb and conduct water and mineral nutrients.

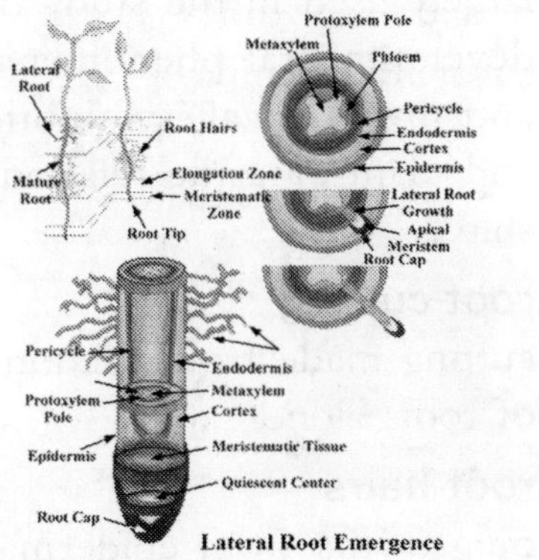

■ root apex

the apical meristem of a root; very similar to the shoot apical meristem in that it forms the three meristematic areas: the protoderm (developing into the epidermis); the procambium (which develops into the stele); and the growth meristem (which forms the cortex).

■ **root cap**

a thimblelike mass of cells covering and protecting the apical meristem of a root.

■ **root culture**

the culture of isolated root tips of apical or lateral origin to produce *in vitro* root systems with indeterminate growth habits. Root culture was among the first kinds of plant tissue cultures, and is still largely used in the study of developmental phenomena, and mycorrhizal, symbiotic and plant-parasitic relationships.

■ **root cutting**

cutting made from sections of roots alone.

■ **root hairs**

outgrowths from epidermal cell walls of the root specialised for water and nutrient absorption.

■ **root nodule**

a small round mass of cells that is located on the roots of plants and contains nitrogen-fixing bacteria.

Bean Root Nodules

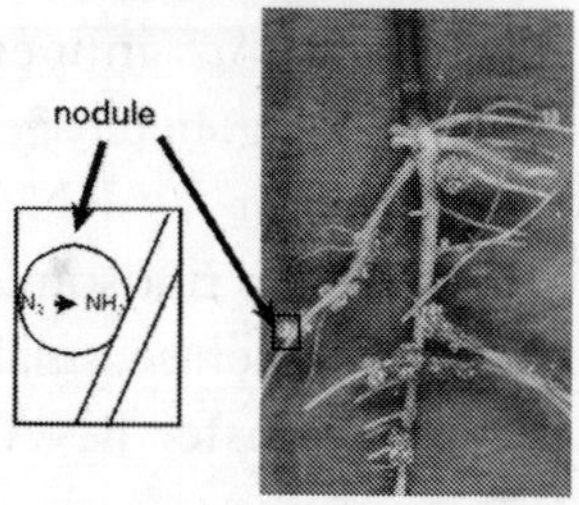

Inside nodules the bacteria fix nitrogen (i.e., convert N_2 into ammonia).

■ **root tuber**

thickened root that stores carbohydrates.

■ **root zone**

the volume of soil or growing medium containing the roots of a plant. In soil science, the depth of the soil profile in which roots are normally found.

■ **rootstock**

the trunk or root material to which buds or scions are inserted in grafting. *See* stock.

■ **rotary shaker**

rotating apparatus with a platform on which, containers can be shaken, such as Erlenmeyer flasks containing cells in liquid nutrient medium.

■ **rRNA**

the RNA molecules which are essential structural and functional components of ribosomes, the organelles responsible for protein synthesis. The different rRNA molecules are known by their sedimentation (Svedberg; symbol S) values. *E. coli* ribosomes contain one 16S rRNA molecule (1541 nucleotides long) in the same (small) sub-unit and a 23S rRNA (2904 nucleotides) and a 5S rRNA (120 nucleotides) in the large sub-unit. These three rRNA molecules are synthesised as part of a large precursor molecule which also contains the sequences of a number of tRNAs. Special processing enzymes cleave this large precursor to generate the functional moieties. *See* RNA.

■ **ruminant animals**

animals having a rumen - a large digestive vat in which fibrous plant material is partially broken down by microbial fermentation, prior to digestion in a "true" stomach (the abomasum). There are also two other stomachs - the reticulum and the omasum. Typical ruminants are cattle and sheep.

■ **runner**

a lateral stem that grows horizontally along the ground surface and gives rise to new plants either from axillary or terminal buds. *cf* stolon.

■ **rust**

a generic descriptor for various plant diseases, especially those caused by a group of parasitic fungi of the phylum *Basidiomycota*, that attack the leaves and stems of crops.

■ **S phase**

the phase in the cell cycle during which DNA synthesis occurs.

■ **S1 nuclease**

an enzyme that specifically degrades RNA or single-stranded DNA to 5´ mononucleotides. Purified from the filamentous fungus *Aspergillus oryzae,* S_1 nuclease is used in assessing the extent of a hybridisation reaction by removing unpaired regions. It is also used to remove the sticky ends of restriction fragments. In S_1 mapping, the coding region of a gene is detected by performing mRNA-DNA hybridisation and removing unpaired DNA with S_1 nuclease.

■ **saccharifaction**

following liquefaction, the hydrolysis of polysaccharides by glucoamylase to maltose and glucose.

■ **salmonella**

a genus of rod-shaped, Gram-negative bacteria that are a common cause of food poisoning.

■ **salt tolerance**

the ability to withstand a concentration of sodium (Na^+ ion), or of any other salt, in the soil (or in culture), which is damaging or lethal to other plants. Breeding and selection for increased tolerance and resistance in crop plants is of great current interest.

■ **sap**

fluid content of the xylem and phloem cells of plants. Fluid c ntents of the vacuole are ref rred to as cell sap.

■ **saprophyte**

a vegetable organism that derives its nutriment from decaying organic matter.

■ **satellite DNA**

that portion of the DNA in plant and animal cells consisting of highly repetitious s quences (millions of copies) typically in the range from 5 to 500 bases. Thousands of copies occur tandemly (end-on-end) at each of many sites. It can be isolated from the rest of the DNA by density gradient centrifugation.

■ **satellite RNA**

a small, self-splicing RNA molecule that accompanies several plant viruses, including tobacco ringspot virus. a.k.a. viroid.

■ **saturates**

noun: oils containing saturated fatty acids.

■ **SCA**

antibody-binding domains in which the two chains are produced by a gene and linked by a short peptide.

■ **scaffold**

the central core structure of condensed chromosomes. The scaffold is composed of non-histone chromosomal proteins.

■ **scale up**

conversion of a process, such as fermentation of a micro-organism, from a small scale to a larger scale.

■ **scanning electron microscope**

an electron-beam-based microscope used to examine, in a three dimensional screen image, the surface structure of prepared specimens.

■ **scarification**

the chemical or physical treatment given to some seeds (where the seed coats are very hard or contain germination inhibitors) in order to break or weaken the seed coat sufficiently to permit germination.

■ **scientific name**

a unique identifier consisting of a genus and a species name (the specific epithet) in Latin, assigned to each recognised and described species of organism. Based on the Linnean system of classification. Susceptible to considerable blurring at the edges because, as Darwin came to realize, nature has not packaged living things into neatly discrete entities. *cf* cultivar; pathovar.

■ **scion**

the twig or bud to be grafted onto another plant, the root stock, in a budding or grafting operation.

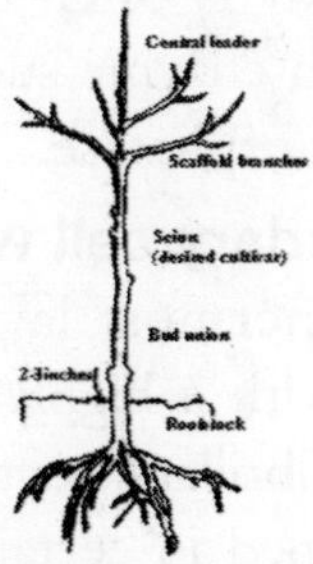

■ **scrapie**

a disease of sheep; a spongiform encephalopathy *See* **proteinaceous infectious particle.**

■ **screen**

to separate by exclusion or collection on the basis of a set of criteria (biochemical, anatomical, physiological, etc.). Screening is often applied to the process of selection for specific purposes, such as disease resistance or improved agronomic qualities in plants, improved performance in animals, specific enzyme properties in micro-organisms, etc.

■ SDS
Sodium dodecyl sulphate.

■ secondary antibody
in an ELISA or other immunological assay system, the antibody that binds to the primary antibody. The secondary antibody is often conjugated with an enzyme such as alkaline phosphatase.

■ secondary cell wall
the innermost layer of cell wall, with a highly organised microfibrillar structure, which is formed in certain cells after cell elongation has ceased. It gives rigidity to the cells.

■ secondary growth
type of growth characterised by an increase in thickness of stem and root and resulting from formation of secondary vascular tissues by the vascular cambium.

■ secondary immune response
the rapid immune response that occurs during the second (and subsequent) encounters of the immune system of a mammal with a specific antigen. *cf* primary immune response.

■ secondary messenger
a chemical compound within a cell that is responsible for initiating the response to a signal from a chemical messenger (such as a hormone) that cannot enter the target cell itself.

■ secondary metabolite
a compound that is not necessary for growth or maintenance of cellular functions but is synthesised, generally, for the protection of a cell or micro-organism, during the stationary phase of the growth cycle.

■ secondary phloem
phloem tissue formed by the vascular cambium during secondary growth in a vascular plant.

■ secondary plant products
metabolic products not having a known functional or structural use in plant cells. They have been extracted from plant tissue cultures for pharmaceutical and food processing purposes (e.g., essential oils, food additives, flavours).

■ secondary root
a branch or lateral root.

■ secondary thickening
deposition of secondary cell wall materials which result in an increase in thickness in stems and roots.

■ secondary vascular tissue
vascular tissue (xylem and phloem) formed by the vascular cambium during secondary growth in a vascular plant.

■ secondary xylem
xylem tissue formed by the vascular cambium during secondary growth in a vascular plant.

■ secretion
the passage of a molecule from the inside of a cell through a membrane into the periplasmic space, or the extracellular medium.

■ seed
botanically, a seed is the matured ovule without accessory parts. Colloquially, a seed is anything which may be sown; i.e., seed potatoes (which are vegetative tubers); seed of corn, sunflower, etc.

■ seed storage proteins
proteins accumulated in large amounts in seeds not because of their enzymatic or structural properties but simply as a convenient source of amino acid for use when the seed germinates. They are of interest to biotechnologists:

■ segment-polarity gene
a gene that functions to define the anterior and posterior components of body segments in *Drosophila*.

■ segregant
a hybrid resulting from the crossing of two genetically unlike individuals.

■ segregation
the separation of the two members of a chromosome pair from each other at meiosis; the result is seen as the separation of alleles from each other in the gametes of heterozygotes; the occurrence of different phenotypes among offspring, resulting from chromosome or allele separation in their heterozygous parents. Mendel's first principle of inheritance (the Law of Segregation) predicts

that heterozygotes will produce equal numbers of gametes containing each allele.

■ selectable

having a gene product that, when present, enables a researcher to identify and preferentially propagate a particular organism or cell type. *See* reporter gene.

■ selectable marker

a gene whose expression allows the identification of:

1. A specific trait or gene in an organism.
2. Cells that have been transformed or transfected with a vector containing the marker gene. *See* b-lactamase.

■ selection

1. differential survival and reproduction phenotypes.
2. a system for either isolating or identifying specific organisms in a mixed culture.

■ selection coefficient

the proportion by which the fitness of a genotype is less than the fitness of a standard genotype, which is usually the genotype with the highest fitness. In general, relative fitness = 1 - *s*.

■ selection culture

a selection based on difference(s) in environmental conditions or in culture medium composition, such that preferred variant cells or cell lines (presumptive or putative mutants) are favoured over other variants or the wild-type.

■ selection differential

the difference between the mean of the individuals selected to be parents and the mean of the overall population.

■ selection pressure

the intensity of selection acting on a population of organisms or cells in culture. Its effectiveness is measured in terms of differential survival and reproduction, and consequently in change in the frequency of alleles in a population.

■ selection response

the difference between the mean of the individuals selected to be parents and the mean of their offspring. Predicted response = heritability (narrow-sense) × selection differential.

selection unit
the minimum number of organisms or cells effective in the screening process.

selective agent
an environmental or chemical agent characterised by its lethal or sub-lethal stress on growing plants, or portion thereof in culture. A selective agent is mainly used when selection of resistant or tolerant individuals is the research aim. *See* single-cell line.

self-fertilisation
the process by which pollen of a given plant fertilises the ovules of the same plant. Plants fertilised in this way are said to have been selfed. An analogous process occurs in some animals, such as nematodes and molluscs.

self-incompatibility
in plants, the inability of the pollen to fertilise ovules (female gametes) of the same plant.

self-pollination
pollen of a plant is transferred to the female part of the same plant or another plant with the same genetic makeup. Opposite: cross-pollination.

self-replicating elements
extrachromosomal DNA elements that have origins of replication for the initiation of their own DNA synthesis.

semi-conservative replication
during DNA duplication, each strand of a parent DNA molecule is a template for the synthesis of its new complementary strand. Thus, one half of a pre-existing DNA molecule is conserved during each round of replication. *cf* replisome.

semi-continuous culture
cells in an actively dividing state are maintained in culture by periodically draining off the medium and replenishing it with fresh medium.

semi-permeable membrane
a cell or plasma membrane that is partially permeable; certain ions or molecules (water, solvents) can pass through it but others cannot (such as certain solutes).

■ **semi-solid**
gelled but not firmly so; small amounts of a gelling agent are used to obtain a semi-solid m dium; called also semi-liquid.

■ **semi-sterility**
a condition of only partial fertility in plant zygotes; usually associated with translocations.

■ **senescence**
the last stage in the post-embryonic development of multicellular organisms, during which loss of functions and degradation of biological components occur. A physiological ageing process in which cells and tissues deteriorate and finally die.

■ **sense RNA**
a primary transcript (RNA) that contains a coding region (contiguous sequence of codons) that is translated to produce a polypeptide.

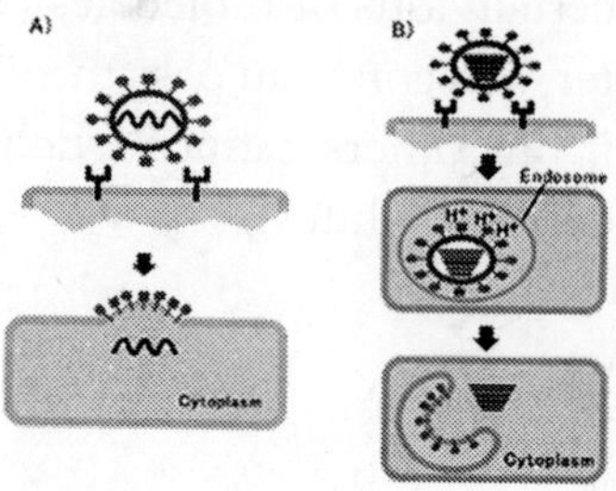

■ **sensitivity**
for diagnostic tests, the smallest amount of the target molecule that the assay can detect.

■ **sepsis**
destruction of tissue by pathogenic micro-organisms or their toxins, especially through infection of a wound. *cf* aseptic; axenic; sterile.

■ **septate**
divided by cross walls into cells or compartments.

■ **septum**
any dividing wall or partition; frequently a cross wall in a fungal or algal filament.

■ **sequence hypothesis**
Francis Crick's seminal concept that genetic information exists as a linear DNA code; DNA and protein sequence are collinear.

■ **sequence-tagged site**
short, unique DNA sequence (usually 200 to 500 bp) that, by being able to be amplified by PCR, is uniquely "tagged" to the site on the chromosome from which it was amplified.

■ **sequencing**
the determination of the order of nucleotides in a DNA or RNA molecule, or that of amino acids in a polypeptide chain. *See* DNA sequencing.

■ **serial divisions**
splitting at about monthly intervals of excised shoot-tip material growing on culture medium, in order to induce additional plantlets.

■ **serial float culture**
a technique of floating anthers on liquid medium developed by Sunderland. Anther dehiscence, pollen release and development occur at intervals of several days, and in different nutrient media.

■ **serology**
the study of serum reactions between an antigen and its antibody. Serology is mainly used to identify and distinguish between antigens, such as those specific to micro-organisms or viruses. Serology is also employed as an indicator technique to assay plants suspected of being virus-infected.

■ **serum albumin**
a globular protein obtained from blood and body fluids.

■ **sewage treatment**
sewage treatment is one of the most widespread biotechnological processes in Western societies, to deal with the huge amounts of human and animal waste that such societies produce. Sewage treatment methods vary widely, but all have a biological basis to break down the organic material in sewage and convert it into something that can be safely discharged into the environment (usually rivers or seas).

■ **sex chromosomes**
chromosomes that are connected with the determination of sex: X and Y chromosomes in mammals; W and Z chromosomes in birds.

■ **sex determination**
the method by which the distinction between males and females is established in a species.

■ **sex factor**
a bacterial episome (e.g., the F plasmid in *E. coli*) that enables

the cell to be a donor of genetic material. The sex factor may be propagated in the cytoplasm, or it may be integrated into the bacterial chromosome.

■ **sex hormones**

steroid hormones that control sexual development.

■ **sex linkage**

the location of a gene on a sex chromosome, typically on the X chromosome.

■ **sexduction**

the incorporation of bacterial genes into F factors and their subsequent transfer, by conjugation, to a recipient cell.

■ **sexed embryos**

embryos separated according to sex.

■ **sex-influenced dominance**

the tendency for the type of gene action to vary between the sexes within a species. Thus the presence of horns in some breeds of sheep appears to be dominant in males and recessive in females.

■ **sex-limited**

expression of a trait in only one sex; e.g., milk production in mammals; egg production in chickens.

■ **sexual reproduction**

the process where two cells (gametes) fuse to form one fertilised cell or zygote. *cf* asexual reproduction; gamete; hybrid.

■ **shake culture**

an agitated suspension in culture providing adequate aeration for cells in the liquid medium. Usually an Erlenmeyer flask containing the culture is attached to a horizontal or platform shaker, or agitated with a magnetic stirrer.

■ **shaker**

a platform fitted with clips for grasping Erlenmeyer flasks, with set or variable speed control. Shaking speed must be adjusted for gentle and even agitation of suspension cultures. *cf* reciprocating shaker.

■ **shear**

1. the sliding of one layer across another, with deformation and fracturing in the direction parallel to the movement. This term usually refers to the forces that cells are subjected to in a bioreactor or a

mechanical device used for cell breakage.
2. to fragment DNA molecules into smaller pieces. DNA, as a very long and fairly stiff molecule, is very susceptible to hydrodynamic shear forces. Forcing a DNA solution through a hypodermic needle will fragment it into small pieces. The size of the fragments obtained is inversely proportional to the diameter of the needle's bore. The actual sites at which the shear force breaks a DNA molecule are approximately random. Therefore DNA fragments may be generated by random shear and then cloned (by either tailing their ends or using linkers) so as to create a complete gene library of an organism. This method is little used now, having been replaced by the use of partial digests with four-base-pair cutters, such as *Sau*3A, as a means of generating random DNA fragments.

■ Shine-Dalgarno sequence

a conserved sequence in prokaryotic mRNAs that is complementary to a sequence near the 5′ terminus of the 16S ribosomal RNA and is involved in the initiation of translation. *See* ribosomal binding site.

■ shoot

a young branch that grows out from the main stock of a tree, or the young main portion of a plant growing above ground.

■ shoot differentiation

the development of growing points, leaf primordia and finally shoots from a shoot tip, axial bud, or even a callus surface.

■ sieve tube

a tube within the phloem tissue of a plant, and composed of joined sieve elements.

■ sigma factor

the sub-unit of prokaryotic RNA polymerases that is responsible for the initiation of transcription at specific initiation sequences.

■ signal sequence

a segment of about 15 to 30 amino acids at the N terminus of a protein, that enables the protein to be secreted (pass through a cell membrane). The signal sequence is removed as the protein is secreted. Also called signal peptide, leader peptide.

signal transduction
the biochemical events that conduct the signal of a hormone or growth factor from the cell exterior, through the cell membrane, and into the cytoplasm. This involves a number of molecules, including receptors, ligands and messengers.

signal-to-noise ratio
a specifically produced response compared to the response level when no specific stimulus (activity) is present.

silencer
a DNA sequence that helps to reduce or shut off the expression of a nearby gene.

simplicity
for diagnostic tests, the ease with which an assay can be implemented.

SINEs
Short interspersed nuclear elements. Families of short (150 to 300 bp), moderately repetitive elements of eukaryotes, occurring about 100,000 times in a genome. SINES appear to be DNA copies of certain tRNA molecules, created presumably by the unintended action of reverse transcriptase during retroviral infection.

single copy
a gene or DNA sequence which occurs only once per (haploid) genome. Most structural genes, those encoding functional proteins, are single-copy genes.

single-cell line
a culture initiated from a single cell, usually from suspension cultures of single cells or small aggregates plated on solidified medium. The latter may incorporate a selective agent, from which tolerant or resistant individual cell lines or cell clones can be selected. *See* selective agent.

single-cell protein
protein produced by micro-organisms. The dried mass of a pure sample of a protein-rich-micro-organism, which may be used either as feed (for animals) or as a food (for humans).

■ **single-node culture**
culture of separate lateral buds with each carrying a piece of stem tissue.

■ **single-nucleotide polymorphism**
a polymorphism (q.v.) at a particular base site in a coding sequence, e.g., at base 306 in a particular gene, one individual could be heterozygous for A and G: the maternal allele could have an A at this site, while the paternal allele has a G at this site. This type of polymorphism is extensive throughout the genome, and has the great advantage of being detectable without the need for gel electrophoresis, which opens the way for large-scale automation of genotyping.

■ **single-strand-DNA-binding protein**
a protein that coats DNA single strands, keeping them in an extended state.

■ **single-stranded**
a term used to describe nucleic acid molecules consisting of only one polynucleotide chain. The genomes of certain phages, e.g., MI3, are single-stranded DNA molecules; rRNA, mRNA and tRNA are all single-stranded nucleic acids, but they all contain double-stranded regions formed by the intra-strand base-pairing of self-complementary sequences.

■ **sires**
male animals used for breeding.

■ **site-directed mutagenesis**
the introduction of base changes - mutations - into a piece of DNA at a specific site, using recombinant DNA methods.

■ **site-specific**
a term used to describe any process or enzyme which acts at a defined sequence within a DNA or RNA molecule. Type II restriction enzymes are site-specific endonucleases and the recombination systems encoded by some transposons are site-specific, such as is the integration of phage into the *E. coli* chromosome.

■ **site-specific mutagenesis**
a technique to change one or more specific nucleotides

within a cloned gene in order to create an altered form of a protein with one or more specific amino acid changes. a.k.a. oligonucleotide-directed mutagenesis; oligonucleotide-directed site-specific mutagenesis.

■ **six-base cutter**

a type II restriction endonuclease that binds (and subsequently cleaves) DNA at sites that contain a sequence of six nucleotide pairs that is uniquely recognised by that enzyme. Because any sequence of six bases occurs less frequently by chance than any sequence of four bases, six-base cutters cleave less frequently than do four-base cutters. Thus, six-base cutters create larger fragments than four-base cutters.

■ **small nuclear ribonucleoprotein**

a compound comprising small nuclear RNA (q.v.) and nuclear protein, that is heavily involved in the post-transcriptional processing of mRNA, especially the removal of introns. snRNPs are a major component of spliceosomes (q.v.).

■ **small nuclear RNA**

short RNA transcripts of 100-300 bp that associate with proteins to form small nuc ear ribonucleoprotein p rticles (snRNPs) (q.v.); most snRNAs are components of the spliceosomes (q.v.) that excise introns from pre-mRNAs in RNA processing. *See* RNA.

■ **soil amelioration**

the improvement of poor soils, usually using bacteria or fungi. This contrasts with bioremediation, which is the cleaning up of toxins, usually in soils. Amelioration includes breaking down organic matter; forming humus; by solubilising them, making minerals - such as phosphates - in the soil available to plants; fixing nitrogen; and sometimes an element of bioremediation as well.

■ **soilless culture**

tissue culture and hydroponics. Growing plants in nutrient solution without soil.

■ **solid media**
nutrient media that has been solidified, such as by addition of agar.

■ **somaclonal variation**
epigenetic or genetic changes, sometimes expressed as a new trait, resulting from *in vitro* culture of higher plants.

■ **somatic**
referring to vegetative or non-sexual stages of a life-cycle.

■ **somatic cell**
any cell of a multicellular organism that composes the body of that organism but does not produce gametes. *cf* gamete; somatic cell gene therapy.

■ **somatic cell embryogenesis**
embryos are produced either from somatic cells of explants (direct embryogenesis) or by induction on callus formed by explants (indirect embryogenesis). a.k.a. asexual embryogenesis.

■ **somatic cell gene therapy**
the delivery of a gene or genes to a tissue other than reproductive cells of an individual, with the aim of correcting a genetic defect. *cf* somatic cell.

■ **somatic cell variant**
a somatic cell with unique characters not present in the other cells, such as might be selected for in a screening trial that following a mutation event.

■ **somatic embryo**
an organised embryonic structure morphologically similar to a zygotic embryo but initiated from somatic (non-zygotic) cells. Under *in vitro* conditions, somatic embryos go through developmental processes similar to embryos of zygotic origin.

■ **somatic hybridisation**
asexual fusion of protoplasts from somatic cells of genetically different parents.

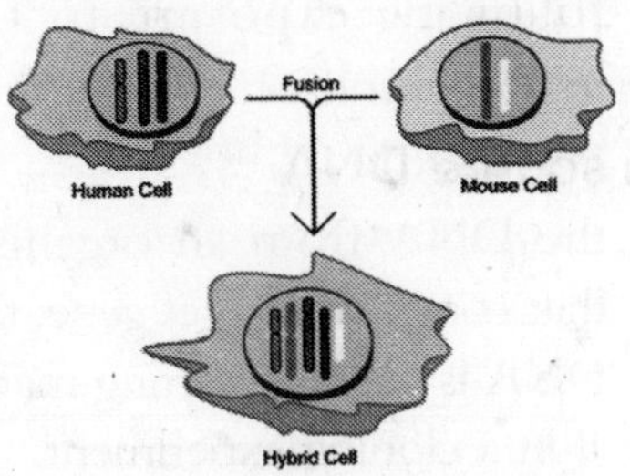

■ **somatic hypermutation**
a high frequency of mutation that occurs in the gene seg-

ments encoding the variable regions of antibodies during the differentiation of B lymphocytes into antibody producing plasma cells.

■ **somatic reduction**

halving of the chromosomal number of somatic cells; a possible method of producing "haploids" from somatic cells and calluses by artificial means.

■ **sonication**

disruption of cells or DNA molecules by high frequency sound waves. a.k.a. ultrasonication.

■ **SOS response**

the synthesis of a whole set of DNA repair, recombination and replication proteins in bacteria containing severely damaged DNA (e.g., following exposure to UV light).

■ **source DNA**

the DNA from an organism that contains a target gene; this DNA is used as starting materi al in a cloning experiment.

■ **source organism**

a bacterium, plant or animal from which DNA is purified and used in a cloning experiment.

■ **southern blot**

a cellulose or nylon membrane to which DNA fragments previously separated by gel electrophoresis have been transferred by capillary action. Named after Ed Southern.

■ **southern blotting**

a technique for transferring denatured DNA molecules that have been separated electrophoretically, from a gel to a matrix (such as a nitrocellulose membrane) on which a hybridisation assay can be performed.

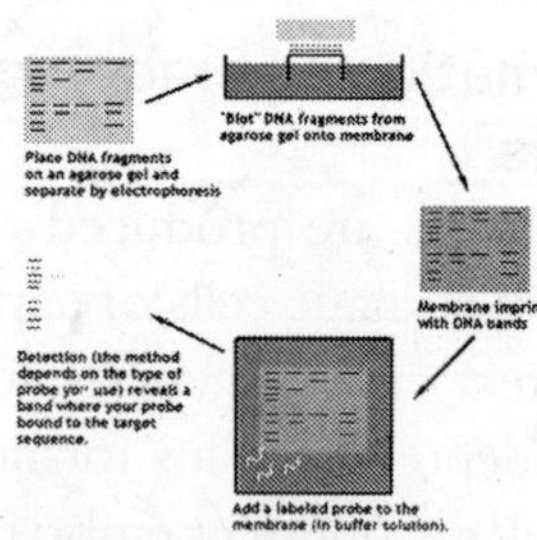

■ **southern hybridisation**

a procedure in which a cloned, labelled segment of DNA is hybridised to DNA restriction fragments on a Southern blot (q.v.).

■ sparger
a device that introduces into a bioreactor air in the form of separate fine streams.

■ specialised
anatomically or physiologically adapted for particular functions or habitats.

■ speciation
the development of one or more species from an existing species.

■ species
a class of potentially interbreeding individuals that are reproductively isolated from other such groups having many characteristics in common. A somewhat arbitrary and sometimes blurred classification; but still quite useful in many situations.

■ specificity
for diagnostic tests, the ability of a probe to react precisely with a specific target molecule.

■ spent medium
after each sub-culture, the medium is discarded because it has been depleted of nutrients, dehydrated or accumulated toxic metabolic products.

■ sperm
abbreviation of spermatozoon.

■ sperm competition
competition between different spermatozoa to reach and fertilise the egg cell of a single female.

■ sperm sexing
the separation of sperm into those bearing an X chromosome and those bearing a Y chromosome, in order to be able to produce, via artificial insemination or *in vitro* fertilisation, animals of a specified sex. Achieved by means of inactivation of X-bearing or Y-bearing sperm via antibodies directed against sex-specific peptides on the surface of sperm cells, or fluorescence-activated cell sorting (FACS) (q.v.), in which sperm that have been pre-treated with a fluorescent dye that binds to DNA are separated according to the quantity of fluorescence detected by a laser beam, based on the principle that X-bear-

ing sperm contain more DNA than Y-bearing sperm.

■ **spermatid**

one of the four cells formed by the meiotic divisions in spermatogenesis. Spermatids become mature spermatozoa (sperm).

■ **spermatocyte**

sperm mother cell. The cell that undergoes two meiotic divisions (spermatogenesis) to form four spermatids; the *primary* spermatocyte before com letion of the first meio ic division; the *secondar* spermatocyte after completion of the first meiotic division.

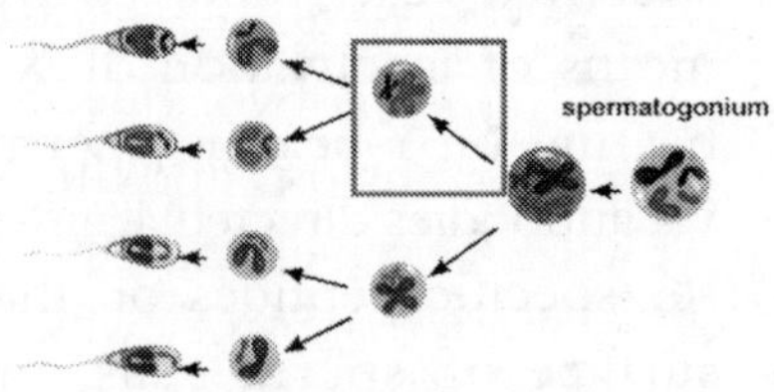

■ **spermatogenesis**

the series of cell divisions in the testis by which maturation of the gametes (sperm) of the male takes place.

■ **spermatogonium**

primordial male germ cell that may divide by mitosis to produce more spermatogonia. A spermatogonium may enter a growth phase and give rise to a primary spermatocyte.

■ **spermatozoon**

the mature, mobile reproductive cell of male animals, produced by the testis.

■ **spharoblast**

nodule of wood which can give rise to adventitious shoots with juvenile characteristics.

■ **spheroplast**

a microbial or plant cell from which most of the cell wall has been removed, usually by enzymic treatment. Strictly, in a spheroplast, some of the wall remains, while in a protoplast the wall has been completely removed. In practice, the two words are often used interchangeably. *See* protoplast.

■ **spike**

an inflorescence in which the main axis is elongated and the flowers are sessile.

■ **spikelet**

the unit of inflorescence in grasses; a small group of grass flowers.

■ **spindle**
in mitosis and meiosis, refers to the spindle-shaped intracellular structure in which the chromosomes move.

■ **spine**
hard, sharp structure on the surface of a plant; usually a modified leaf.

■ **spliceosomes**
organelles responsible for the removal of introns from mRNA by means of splicing (q.v.).

■ **splicing**
during the maturation of eukaryotic mRNA, the process that eliminates intervening intron sequences and covalently joins exon sequences of RNA. *cf* split gene; exon; guide sequence. In recombinant DNA technology, the term refers to the latter of the two processes just described, namely joining fragments of DNA together. *See* gene splicing.

■ **split genes**
in eukaryotes, structural genes are typically divided up (split) by a number of non-coding regions called introns. *cf* exon; guide sequence; splicing.

■ **sporophyll**
a leaf that bears spore producing structures (sporangia).

■ **standard deviation**
a statistical measure of variability in a population of individuals or in a set of data; the square root of the variance.

■ **standard error**
a statistical measure of variation in a population of means, used to indicate how well sample estimates represent population parameters.

■ **starch**
a complex insoluble carbohydrate, consisting of various proportions of two glucose polymers, amylose and amylopectin; the chief food storage substance of plants, which is composed of several hundred hexose sugar units and which easily breaks down on hydrolysis into these separate units.

■ **start codon**
the set of three nucleotides in an mRNA molecule with which the ribosome starts the

process of translation. The start codon sets the reading frame for translation. The most commonly used start codon is AUG, which is decoded as methionine in eukaryotes and as *N*-formylmethionine in prokaryotes. AUG appears to be the only start codon used by eukaryotes, while in bacteria, GUG (valine) may sometimes be employed.

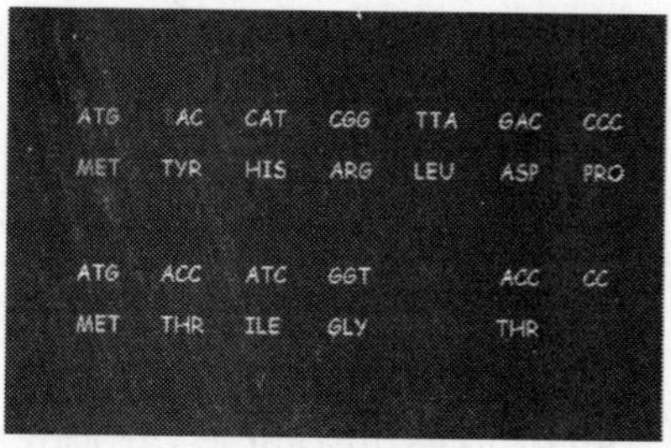

■ **stationary culture**

a culture maintained in the growth chamber with no agitation movement. The antonym is shake culture.

■ **stationary phase**

the plateau of the growth curve after log growth, during which cell number remains constant. New cells are produced at the same rate as older cells die.

■ **statistic**

an estimate based on a sample or samples of a population, providing an indication of the true population parameter.

■ **steady state**

in a continuous fermentation process, the condition when the number of cells that are removed with the outflow is exactly balanced by newly synthesised cells.

■ **stele**

the central cylinder, inside the cortex, of roots and stems of vascular plants. *See* meristele.

■ **stem**

the main body of the above-ground portion of a tree, shrub, herb or other plant; the ascending axis, whether above or below ground, of a plant.

■ **stem cell**

an undifferentiated active somatic cell that undergoes division and gives rise to other stem cells or to cells that differentiate to form specialised cells.

■ **sterilisation**

the act of sterilising.

■ **Steward bottle**
flask developed by Steward for the growth of cells and tissues in a liquid medium, in which they can be periodically submerged during rotation.

■ **sticky ends**
the single-stranded nucleotide sequence left on a restriction fragment by type II restriction enzymes that cut each strand at a separate location. These unpaired regions are available for hybridisation with complementary ends on other fragments during the creation of recombinant DNA. a.k.a. protruding end; overhang; cohesive end.

■ **stigma**
receptive portion of the style, to which pollen adheres.

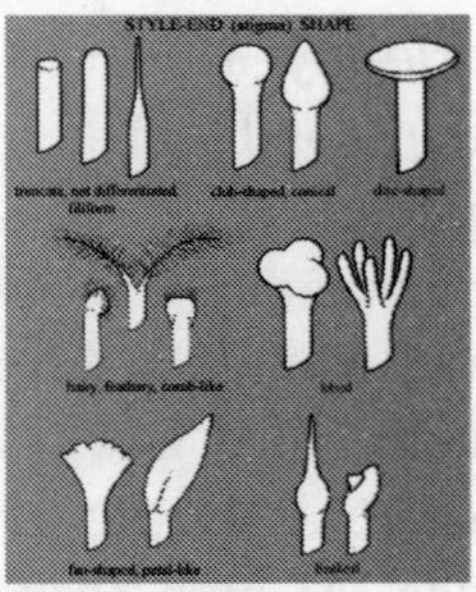

■ **stirred-tank fermenter**
a growth vessel in which cells or micro-organisms are mixed by mechanically-driven impellers.

■ **stock**
the lower portion of a graft.

■ **stock plant**
the source plant from which cuttings or explants are made. Stock plants are usually maintained carefully in an optimum state for (sometimes prolonged) explant use. Preferably they are certified-pathogen-free plants.

■ **stock solutions**
pre-prepared solutions of individual components and used to prepare many different types of media. Certain substances, including Ca and Mg sulphates and phosphates must not be combined until actual medium preparation because insoluble combinations are formed and precipitate.

■ **stolon**
a lateral stem that grows horizontally along the ground surface. The runners of white clover, strawberry and bermuda grass are examples of stolons. *cf* runner.

■ stomatal complex

includes the stoma, together with its guard cells and, when present, the subsidiary cells.

■ stomatal index

= (number of stomata per mm^2 × 100)/(number of stomata per mm^2 + number of epidermal cells per mm^2). This value has been found useful in comparing leaves of different sizes. Relative humidity and light intensity during leaf development affect the value of stomata index.

■ stop codon

a set of three nulecotides for which there is no corresponding tRNA molecule to insert an amino acid into the polypeptide chain. Protein synthesis is hence terminated and the completed polypeptide released from the ribosome. Three stop codons are found: UAA (ochre), UAG (amber) and UGA (opal). Mutations which generate any of these three codons in a position which normally contains a codon specifying an amino acid are known as *nonsense mutations*. Stop codons can also be called nonsense codons. *See* chain terminator; nonsense mutation; suppressor.

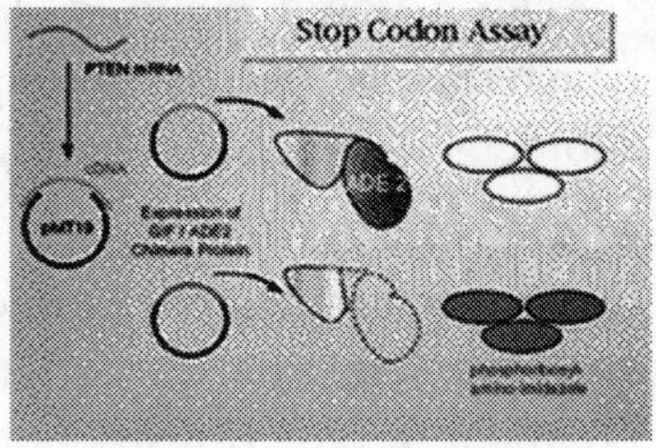

■ strain

a group of individuals from a common origin within a species.

■ strain isolation

isolation of any bacterium, animal or plant from the outside world.

■ stratification

treatment of moist seeds at low temperature (+2° - +4°C) to break dormancy.

■ stringency

reaction conditions - notably temperature, salt concentration(s) and pH - that dictate the annealing of single-stranded DNA/DNA, DNA/RNA and RNA/RNA hybrids. At high stringency, duplexes form only between strands with perfect one-to-one complementarity; lower stringency allows annealing between strands with some degree of mismatch between bases.

■ stringent plasmid

a plasmid that only replicates along with the main bacterial chromosome and is present as a single copy, or at most several copies, per cell. *See* plasmid.

■ stroma

tissue that forms the framework of an organ.

■ structural gene

a DNA sequence that forms the blueprint for the synthesis of a polypeptide.

■ structure-functionalism

the scientific tradition that stresses the relationship between a physical structure and its function, e.g., the related disciplines of anatomy and physiology.

■ style

slender column of tissue that arises from the top of the ovary and through which the pollen tube grows.

■ sub-clone

a method in which smaller DNA fragments are cloned from a large insert which has already been cloned in a vector.

■ symbiont

an organism living in symbiosis with another, dissimilar organism.

■ symbiotic association

an intimate partnership between two organisms, in which the mutual advantages normally outweigh the disadvantages.

■ sympatric speciation

the formation of new species by populations that inhabit the same or overlapping geogra hic regions.

■ symplast

the system of protoplasts in plants, that are interconnected by plasmodesmata.

■ sympodial

a type of plant development in which the terminal bud of the stem stops growing due to either its abortion or its development into a flower or an inflorescence, and the uppermost lateral bud takes over the further axial growth of the stem.

Sympodial Growth Patterns

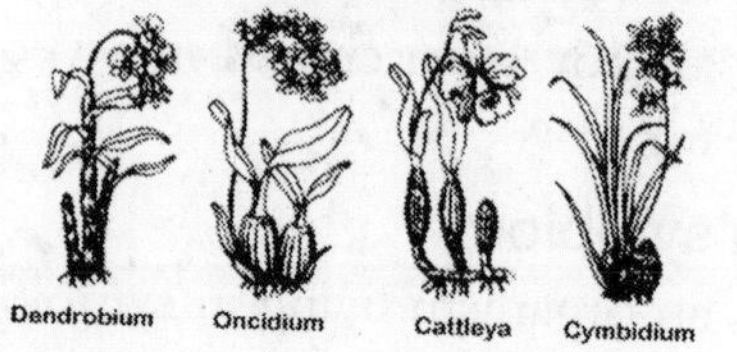

■ synaptonemal complex
a ribbonlike protein structure formed between synapsed homologues at the end of the first meiotic prophase, binding the chromatids along their length and facilitating chromatid exchange.

■ synchronous culture
a culture in which the majority of the cells are dividing at the same time or are at a specific phase of the cell cycle.

■ syncytium
a group of cells in which cytoplasmic continuity is maintained. *cf* acellular.

■ syndrome
a group of symptoms that occur together and represent a particular disease.

■ synergids
the two nuclei within the embryo sac at the upper end in the ovule of the flower, which, with the third (the egg), constitute the egg apparatus.

■ syngamy
fertilisation.

■ synkaryon
a nucleus formed by the fusion of nuclei from two different somatic cells during somatic-cell hybridisation.

■ synteny
the occurrence of two or more loci on the same chromosome, without regard to the distance between them.

■ T cell receptor
an antigen-binding protein that is located on the surfaces of killer T cells and mediates the cellular immune response of mammals. The genes that encode T cell antigens are assembled from gene segments by somatic recombination processes that occur during T lymphocyte differentiation.

■ T cells
lymphocyte that pass through the thymus gland during maturation. Different kinds of T cells play important roles in the immune response, being primarily responsible for the

T cell-mediated response or cellular immune response.

■ T4 DNA ligase

an enzyme from bacteriophage-T4-infected cells, and that catalyses the joining of duplex DNA molecules and repairs nicks in DNA molecules. The enzyme requires that one of the DNA molecules has a 5´-phosphate group and that the other has a free 3´-hydroxyl group.

■ tailing

the *in vitro* addition of the same nucleotide by the enzyme terminal transferase, to the 3´-hydroxyl ends of a duplex DNA molecule. a.k.a. homopolymeric trailing.

■ tandem array

the existence of two or more identical DNA sequences in series, i.e., end -to-end.

■ tank bioreactor

vessel in which fermentation takes place. A tank bioreactor is a vessel in which a microorganism is grown in a large volume of liquid. This contrasts with fibre or mem rane bioreactors and immobilised cell reactors. The l rge majority of bioreactors used in biotechnology are tank bioreactors, and most tank bioreactors are stirred-tank bioreactors, because stirring helps to distribute effectively gas and nutrients to the growing organism.

■ tap root

root system in which the primary root has a much larger diameter than the lateral roots. Opposite: fibrous root.

■ target

for diagnostic tests, the molecule or nucleic acid sequence that is being sought in a sample.

■ target site duplication

a sequence of DNA that is duplicated when a transposable element inserts; usually found at each end the insertion.

■ targeted drug delivery

a method of delivering a drug to the site in the body where it is needed, rather than allowing it to diffuse into many sites.

■ targeting vector

a cloning vector carrying a DNA sequence capable of

participating in a crossing-over event at a specified chromosomal location in the host cell.

■ **TATA box**

a conserved adenine- and thymine-rich promoter sequence located 25-30 bp upstream of a gene, which is the binding site of RNA polymerase. *See* Pribnow box.

■ **tautomeric shift**

the transfer of a hydrogen atom from one position in an organic molecule to another position.

■ **tautomerism**

a type of isomerism in which the two isomes are in equilibrium.

■ **T-cell-mediated (cellular) immune response**

the synthesis of antigen-specific T cell receptors and the development of killer T cells in response to an encounter of immune system cells with a foreign immunogen.

■ **T-DNA**

the segment of DNA in the Ti plasmid of *Agrobacterium tumefaciens* that is transferred to plant cells and inserted into the chromosomes of the plant.

■ **telemeter**

the unique structure found at the end of eukaryotic chromosomes containing specialised sequences of DNA that assures the completion of a cycle of DNA replication.

■ **telomerase**

an enzyme that adds telomeric sequences to the ends of eukaryotic chromosomes.

■ **telophase**

the last stage in each mitotic or meiotic division, in which the chromosomes are assembled at the poles of the division spindle.

■ **TEM**

see transmission electron microscope.

■ **temperate phage**

a phage (virus) that invades but may not destroy (lyse) the host (bacterial cell). However, it may subsequently enter the lytic cycle.

■ **temperature-sensitive mutant**
an organism that can grow at one temperature but not at another.

■ **temperature-sensitive protein**
a protein that is functional at one temperature but loses function at another (usually higher) temperature.

■ **template**
an RNA or single-stranded DNA molecule upon which a complementary nucleotide strand is synthesised. A pattern or mould. DNA stores coded information and acts as a model or template from which information is copied into complementary strands of DNA or transcribed into mRNA.

■ **template strand**
the polynucleotide strand that a polymerase uses for determining the sequence of nucleotides during the synthesis of a new nucleic acid strand.

■ **term finalisation**
repelling movement of the centromeres of bivalents in the diplotene stage of the meiotic prophase, that tends to move the visible chiasmata toward the ends of the bivalents.

■ **terminal bud**
a branch tip, an undeveloped shoot containing rudimentary floral buds or leaves, enclosed within protective bud scales.

■ **terminal transferase**
an enzyme that adds nucleotides to the 3′ terminus of DNA molecules.

■ **termination signal**
in transcription, a nucleotide sequence that specifies RNA chain termination.

■ **terminator**
1. a DNA sequence just downstream of the coding segment of a gene, which is recognised by RNA polymerase as a signal to stop synthesizing mRNA. In prokaryotes, terminators usually have an inverted repeat followed by a short stretch of Us at the very end of the transcribed portion. There may also be sequences beyond the transcribed part of the gene which influence the termination of transcription.

2. a name given to antisense DNA inserted in plants to make impossible the use of a second generation of seed by a farmer.

■ **terminator region**
a DNA sequence that signals the end of transcription.

■ **test tube**
tube in which cells, tissues, etc., can be cultured.

■ **testcross**
backcross to the recessive parental type, or a cross between genetically unknown individuals and a fully recessive tester to determine whether an individual in question is heterozygous or homozygous for a certain allele. It is also used as a test for linkage, i.e., to estimate recombination fraction.

■ **tetracycline**
an antibiotic that interferes with protein synthesis in prokaryotes.

■ **tetrad**
1. the four cells arising from the second meiotic division in plants (pollen tetrads) or fungi (ascospores).

2. the quadruple group of chromatids that is formed by the association of duplicated homologous chromosomes during synapsis in meiosis I. a.k.a. quadrivalent.

■ **tetraploid**
an organism whose cells contain four haploid (4*x*) sets of chromosomes.

■ **tetrasomic**
pertaining to a nucleus or an organism with four members of one of its chromosomes, whereas the remainder of its chromosome complement is diploid. Chromosome formula: 2n + 2.

■ **tetratype**
in fungi, a tetrad of spores that contains four different types; e.g., *AB*, *aB*, *Ab* and *ab*.

■ **thallus**
plant body without true roots, stems, or leaves.

■ therapeutic agent

a compound that is used for the treatment of a disease or for improving the well-being of an organism. a.k.a. pharmaceutical agent; drug; protein drug.

■ thermic shock

exposure to reduced or increased temperature for several days.

■ thermolabile

not heatproof, e.g., a substance which disintegrates or is unstable upon heating.

■ thermophile

an organism which grows at a higher temperature than most other organisms. In general, a wide range of bacteria, fungi and simple plants and animals can grow at temperature up to 50°C; thermophiles are considered to be organisms which can grow at above 50°C. They can be classified according to their optimal growth temperature, into simple thermophiles (50-65°C), thermophiles (65-85°C), and extreme thermophiles (>85°C). Thermophiles and extreme thermophiles are usually found growing in very hot places, such as hot springs and geysers, smoker vents on the sea floor, and domestic hot water pipes.

■ thermosensitivity

loss of activity of a protein at high temperature.

■ thermostability

retention of activity at high temperature.

■ thermotherapy

technique mainly used for virus or mycoplasma eliminat on. Plants are exposed to ele ated temperatures as a treat ment. Thermotherapy is - used alone or in combination w th meristem culture or meri tem tip culture.

■ thinning

1. removal of older stems to promote new growth.

2. removal of excess fruits to improve the size and quality of the remaining fruits.

3. removal of seedlings spaced too closely for optimum growth.

■ **thymidine**

a nucleoside present in DNA but absent in RNA.

■ **thymidine kinase**

an enzyme that allows a cell to utilize an alternate metabolic pathway for incorporating thymidine into DNA. Used as a selectable marker to identify transfected eukaryotic cells.

■ **thymine**

a pyrimidine base found in DNA. The other three organic bases - adenine, cytosine, and guanine - are found in both RNA and DNA; in RNA, thymine is replaced by uracil.

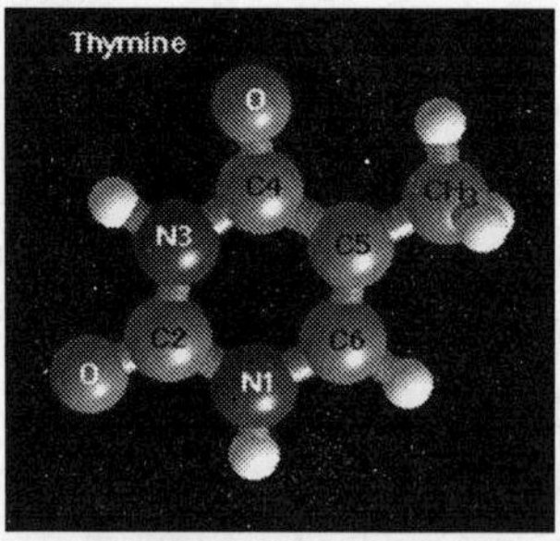

■ **Ti plasmid**

tumour-inducing plasmid. A giant plasmid of *Agrobacterium tumefaciens* that is responsible for the induction of tumours in infected plants. Ti plasmids are used as vectors to introduce foreign DNA into plant cells.

■ **tissue**

a group of cells of similar structure which sometimes performs a special function.

■ **tissue culture**

a general term used to describe the culture of cells, tissues or organs in a nutrient medium under sterile conditions.

■ **tolerance**

a form of genetic resistance in which an organism attacked or affected by a disease pathogen (or pest) exhibits less reduction in yield or performance in comparison with members of other affected cultivars or breeds.

■ **tonoplast**

the cytoplasmic membrane bordering the vacuole, with a role in regulating the pressure exerted by the cell sap.

■ **topo-isomerase**

an enzyme that introduces or removes supercoils from DNA.

■ **Torr**

an obsolete unit of pressure equal to that exerted by a column of mercury 1mm high at 0°C and standard gravity (1mm Hg); named after Evangelista Torricelli (1608-1647), the inventor of the mercury barometer. 1 Torr = 1/760 atm = 133.322 Pa.

■ **totipotency**

having the potentiality of forming all the types of cells in the body. The property of somatic cells to be induced to undergo regeneration. The diploid zygote formed at fertilisation is a single cell which is capable of division and differentiation to give rise to the total range of cell types found in the adult organism.

■ **totipotent cell**

an undifferentiated cell (or nucleus), such as a blastomere, that, when isolated or suitably transplanted, can develop into a complete embryo.

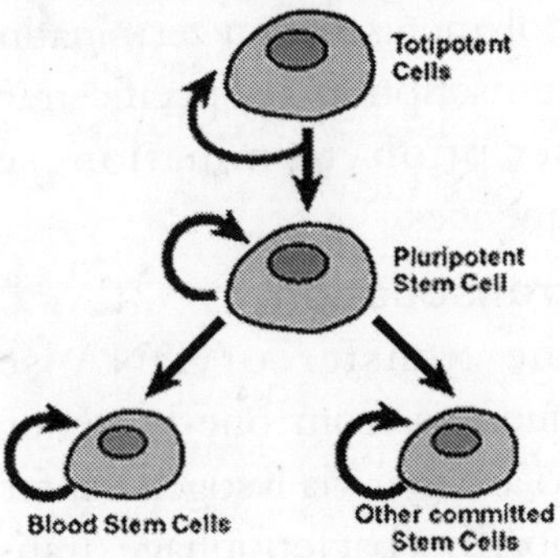

■ **toxicity**

negative effect of a compound, as shown by altered morphology or physiology. It is meaningful only when the effect itself is also described, such as changes in the rate of cell growth, cell death, etc.

■ **toxin**

a compound produced by an organism and poisonous to plants or animals.

■ **tracer**

an added or injected substance that can be followed within a reaction or an organism, such as radioactive isotopes and certain dyes.

■ **tracheid**

an elongated, tapering xylem cell, with lignified pitted walls, and adapted for conduction and support. Found in conifers, ferns and related plants.

Xylem tracheid (from macerated plant tissues)

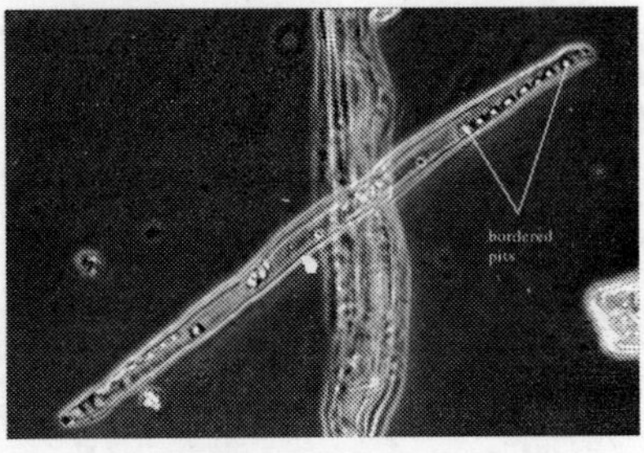

■ **trans heterozygote**

a double heterozygote that contains two mutations arranged in the *trans* configuration.

■ **trans-acting**

a term describing substances that are diffusable and that can affect spatially separated entities within cells.

■ **transcapsidation**

the partial or full coating of the nucleic acid of one virus with a coat protein of a differing virus. *See* coat protein.

■ **transcript**

an RNA molecule that has been synthesised from a specific DNA template. In eukaryotes, the primary transcript produced by RNA polymerase must often be processed or modified in order to form the mature, functional mRNA, rRNA or tRNA.

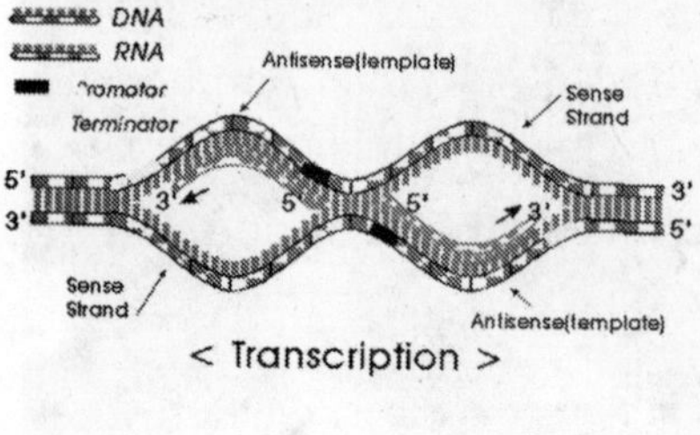

< Transcription >

■ **transcription**

process through which RNA is formed along a DNA template. The enzyme RNA polymerase catalyses the formation of RNA from ribonucleoside triphosphates.

■ **transcription factor**

a protein that regulates the transcription of genes.

■ **transcription unit**

a segment of DNA that contains signals for the initiation and termination of transcription, and is transcribed into one RNA molecule.

■ **transcription vector**

a cloning vector that allows the foreign gene or DNA sequence to be transcribed *in vitro*.

■ **transcriptional antiterminator**

a protein that prevents RNA polymerase from terminating transcription at specific transcription termination sequences.

■ **transduction**

the transfer of DNA sequences from one bacterium to another via lysogenic infection by a bacteriophage (transducing phage). Genetic recombination in bacteria mediated by bacteriophage. Abortive t: Bacterial DNA is injected by a phage into a bacterium, but unable to replicate.

■ **transfection**

the transfer of DNA to an eukaryotic cell.

■ **transfer**

the process of moving cultured tissue or cells to a fresh medium.

■ **transfer RNA**

(tRNA) RNA that transports amino acids to the ribosomes, where the amino acids are assembled into proteins.

■ **transferase**

enzyme that catalyses the transfer of a group of atoms from one molecule to another.

■ **transformant**

in prokaryotes, a cell that has been genetically altered through the uptake of foreign DNA. In higher eukaryotes, a cultured cell that has acquired a malignant phenotype. *See* transformation.

■ **transformation**

1. the uptake and establishment of DNA in a bacterium or yeast cell, in which the introduced DNA often changes the phenotype of the recipient organism.

2. conversion by various means of animal cells in tissue culture from controlled to uncontrolled cell growth. Typically through infection by a tumour virus or transfection with an oncogene.

■ **transformation efficiency**

the number of cells that take up foreign DNA as a function of the amount of added DNA; expressed as transformants per microgram of added DNA. *See* transformation.

■ **transformation frequency**

the fraction of a cell population that takes up foreign DNA; expressed as the number of transformed cells divided by the total number of cells in a population.

■ **transforming oncogene**

a gene that upon transfection converts a previously immortalized cell to the malignant phenotype.

■ **transgene**

a gene from one genome that has been incorporated into the genome of another organism. Often refers to a gene that has been introduced into a multicellular organism.

■ **transgenesis**

the introduction of a gene or genes into animal or plant cells, which leads to the transmission of the input gene (transgene) to successive generations. *cf* transfection.

■ **transgenic**

an organism in which a foreign gene (a transgene) is incorporated into its genome. The transgene is present in both somatic and germ cells, is expressed in one or more tissues, and is inherited by offspring in a Mendelian fashion.

■ **transgressive variation**

the appearance in the F_2 (or later) generation of individuals showing more extreme development of a trait than either of the original parents.

■ **transient**

of short duration.

■ **transition**

the substitution in DNA or RNA of one purine by another purine, or of one pyrimidine by another pyrimidine.

■ **transition stage**

the integration period of juvenile and reproductive stages of growth.

■ **transition-state intermediate**

in a chemical reaction, an unstable and high-energy configuration assumed by reactants on the way to making products. Enzymes are thought to bind and stabilise the transition state, thus lowering the energy of activation needed to drive the reaction to completion.

■ **translation**

the process of polypeptide synthesis in which the amino acid sequence is determined by mRNA, mediated by tRNA molecules, and carried out on ribosomes.

■ **translocation**

1. the movement of nutrients or products of metabolism from one location to another.

2. change in position of a segment of a chromosome to another, non-homologous chromosome.

■ transmission electron microscope

a microscope which uses an electron beam to obtain images of objects, with a much gr ater resolving power than a light microscope. *See* electron microscope.

■ transplant

1.noun: a plant grown in a cold frame, greenhouse, tissue culture or indoors for later planting outdoors.

2. to dig up and move a plant to another location.

■ transposable genetic element

a DNA element that can move from one location in the genome to another. *See* transposon.

■ transposase

an enzyme encoded by a transposon gene and that catalyses the movement of a DNA sequence to a different site in a DNA molecule, by catalysing the excision of the transposon from one site and its insertion into a new chromosomal site.

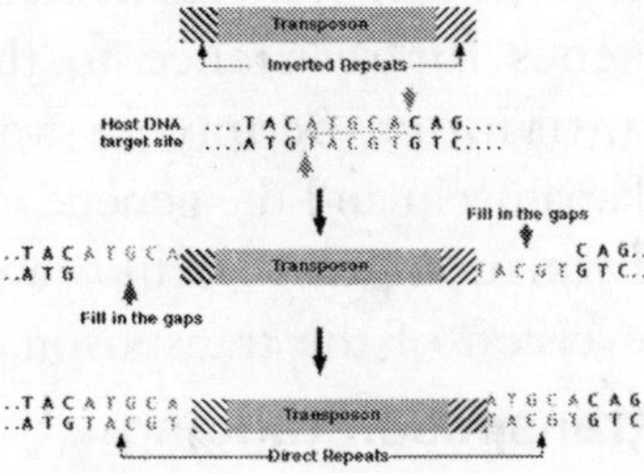

■ transposition

the process whereby a transposon or insertion sequence inserts itself into a new site on the same or another DNA molecule. The exact mechanism is not fully understood and different transposons may transpose by different mechanisms. Transposition in bacteria does not require extensive DNA homology between the transposon and the target DNA. The phenomenon is therefore described as illegitimate recombination.

■ transposon

a transposable or movable genetic element. A relatively small DNA segment that has the ability to move (mobile genetic element) from one chromosomal position to another, e.g., Tn 5 is a bacterial transposon that carries the genes for resistance to the antibiotics neomycin and kanamycin and the genetic information for insertion and excision of the transposon.

■ **transposon tagging**

the insertion of a transposable element into or nearby a gene, thereby marking that gene with a known DNA sequence.

■ **transversion**

the substitution of a purine for a pyrimidine, or a pyrimidine for a purine.

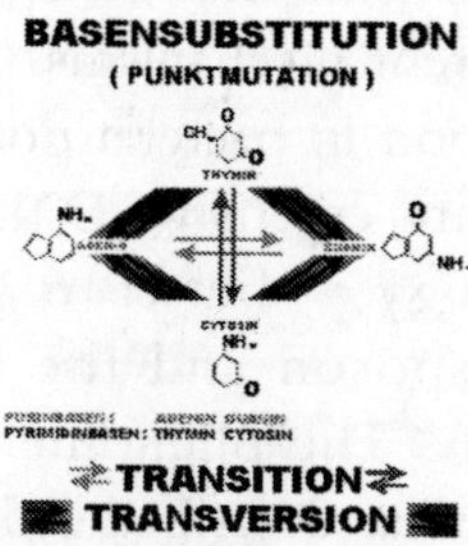

■ **tribrid protein**

a fusion protein that has three segments, each encoded by parts of different genes.

■ **trichome**

a short filament of cells. *See* hair.

■ **tri-hybrid**

the offspring from homozygous parents differing in three pairs of genes.

■ **tri-nucleotide repeats**

tandem repeats of three nucleotides that are present in many genes. In several cases, these trinucleotide repeats have undergone expansions in copy number, and that has resulted in inherited diseases.

■ **tripartite mating**

a process in which conjugation is used to transfer a plasmid vector to a target cell when the plasmid vector is not self-mobilisable. When (1) cells that have a plasmid with conjugative and mobilising functions are mixed with (2) cells that carry the plasmid vector and (3) target cells, mobilising plasmids enter the cells with the plasmid vector and mobilise the plasmid vector to enter into the target cells. Following tripartite mating, the target cells with the plasmid vector are separated from the other cell types in the mixture by various selection procedures.

■ triplet

a sequence of three nucleotides of DNA which specifies an amino acid. The elucidation of the genetic code involved the binding of charged tRNA species to chemically synthesised ribonucleotide triplets.

■ triploid

a cell or organism containing three times the haploid number of chromosomes.

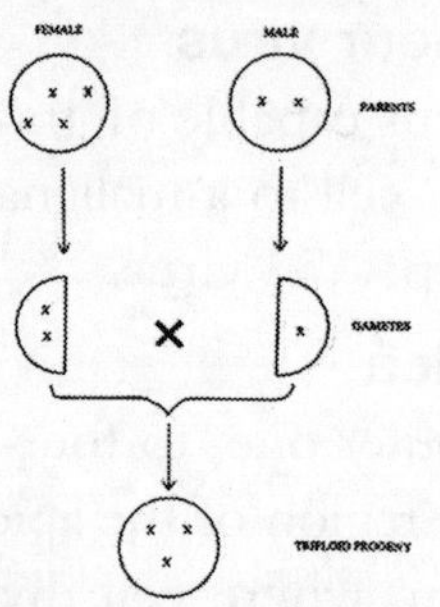

■ trisomy

an otherwise diploid cell or organism that has an extra chromosome of one pair (chromosome formula: 2n + 1). *cf* disomy; monosomy.

■ triticale

the species formed by crossing wheat and rye.

■ tRNA

transfer RNA .The class of small RNA molecules that transfer amino acids to the ribosome during protein synthesis. Transfer RNA molecules are folded into a 'clover-leaf' secondary structure by intrastrand base pairing. The anticodon loop contains a nucleotide-triplet complementary to a specific codon within the mRNA molecule. Each tRNA is 'charged' with the correct amino acid molecule, via its 3′ adenosine moiety, by an enzyme called aminoacyl-tRNA synthetase.

■ tropism

an involuntary plant response to a stimulus, in which a bending, turning or growth occurs, such as phototropism, geotropism or hydrotropism. The response may be positive (towards) or negative (away from) to the stimulus.

■ true-to-type

applied to a plant or propagation source, this term denotes correct cultivar identification and lack of variation in productivity or performance. Verification is determined visually by an expert or through biochemical, serological or other means.

■ trypsin

a proteolytic enzyme that hydrolyzes peptide bonds on the carboxyl side of the amino acids arginine and lysine.

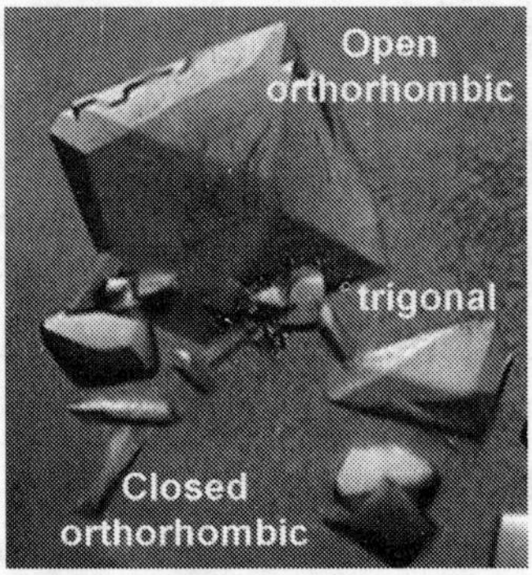

■ trypsin inhibitors

substances inactivating the enzyme trypsin, which is needed for digestion of peptides.

■ tuber

food-storing modified roots in plants like potato.

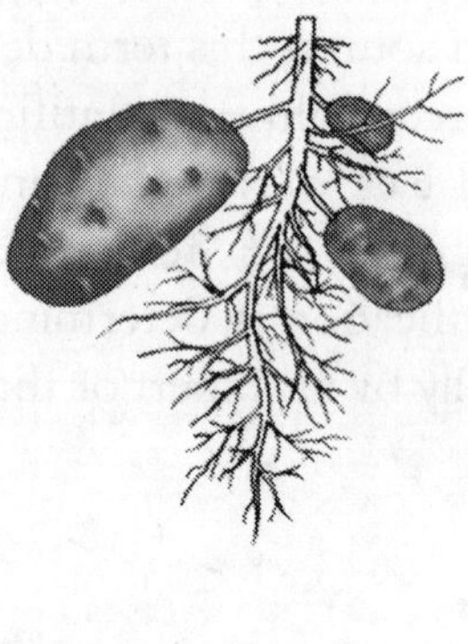

■ tubulin

the major protein component of the microtubules of eukaryotic cells.

■ tumble tube

a glass tube mainly used *in vitro* to agitate and consequently aerate suspension cultures. The tube, which is commonly attached to a slowly revolving platform, is closed at both ends, with a side-neck opening.

■ tumour virus

a virus capable of transforming a cell to a malignant phenotype. *See* virus.

■ tunica

an outer one- to four-cell layered region of the apical meristem, where cell division is anticlinal, i.e., perpendicular to the surface. *See* apical meristem.

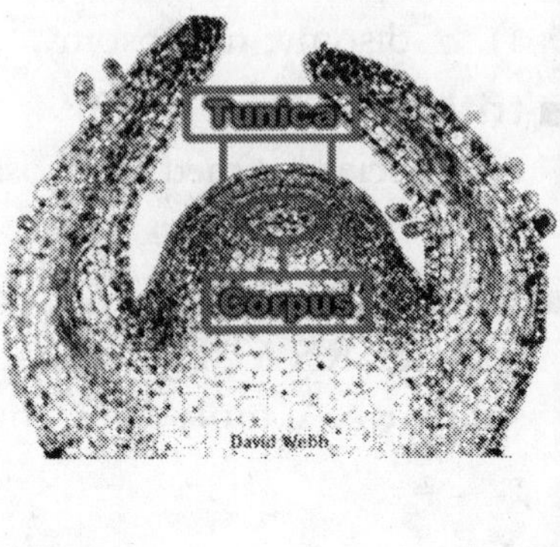

■ turbidostat
an open continuous culture in which a pre-selected biomass density is uniformly maintained by automatic removal of excess cells. The fresh medium flows in response to an increase in the turbidity (usually cell density) of the culture.

■ turgid
swollen, distended; referring to a cell that is firm due to water uptake.

■ turgor pressure
the pressure within the cell resulting from the absorption of water into the vacuole and the imbibition of water by the protoplasm.

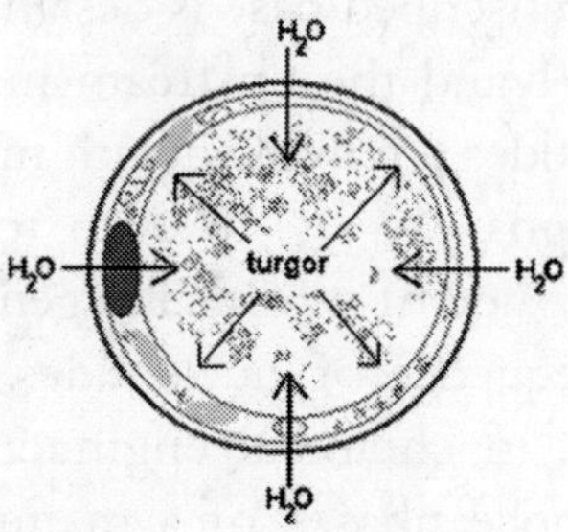

■ turion
an underground bud or shoot from which an aerial stem arises. *cf* sucker.

■ twins
two individuals originating from the same zygote.

■ U
the uracil residue in RNA.

■ ultraviolet light
the portion of the electromagnetic spectrum with wavelengths from about 100 to 400 nm; between ionising radiation (X-rays) and visible light. UV is absorbed by DNA and is highly mutagenic to unicellular organisms and to the epidermal cells of multicellular organisms. UV light is used in tissue culture for its mutagenic and bactericidal properties.

■ undefined
a medium or substance added to medium in which not all of the constituents or their concentrations are chemically defined, such as media contaning coconut milk, malt extract, casein hydrolysate, fish emulsion or other complex compounds. *cf* organic complex.

■ understock
host plant for a grafted scion, a branch or shoot from an-

other plant; an understock may be a fully grown tree or a stump with a living root system.

■ **undifferentiated**

in a meristematic state or resembling a meristem; lacking the specialised or differential gene expression characteristic of specialised cells.

■ **unequal crossing over**

crossing over between repeated DNA sequences that have paired out of register, creating duplicated and deficient products.

■ **unicellular**

describing tissues, organs or organisms consisting of a single cell.

■ **unisexual**

describing animals and plants possessing either male or female reproductive organs, but not both.

■ **univalent**

an unpaired chromosome at meiosis.

■ **universal donor cells**

cells that, after introduction into a recipient, will not induce an immune response that leads to their rejection.

■ **universality**

referring to the genetic code, the codons have the same meaning, with minor exceptions, in virtually all species.

■ **unorganised growth**

in vitro formation of tissues with few differentiated cell types and no recognisable structure; many call are unorganised. *cf* organised growth.

■ **upstream**

1. in molecular biology, the stretch of DNA base pairs that lie in the 5´ direction from the site of initiation of transcription. Usually the first transcribed base is designated +1 and the upstream nucleotides are marked with minus signs, e.g., -1, -10. Also, to the 5´ side of a particular gene or sequence of nucleotides.

2. in chemical engineering, those phases of a manufacturing process that precede the biotransformation step. Refers to the preparation of raw materials for a fermenta-

tion process. Also called upstream processing.

■ **uracil**

a pyrimidine base found in RNA but not in DNA. In DNA, uracil is replaced by thymine.

■ **V regions**

variable regions in antibodies.

■ **v/v**

on a volume per volume basis; the percent of the volume of a constituent in 100 units of volume, e.g., (ml/100 ml) × 100.

■ **vaccine**

a preparation of dead or weakened pathogens, or of derived antigenic determinants, that is used to induce formation of antibodies or immunity against the pathogen.

■ **vaccinia**

the cowpox virus used to vaccinate against smallpox and, experimentally, as a carrier of genes for antigenic determinants cloned from other disease organisms.

■ **vacuole**

a cavity in a plant cell, bounded by a membrane; in which various plant products and by-products are stored.

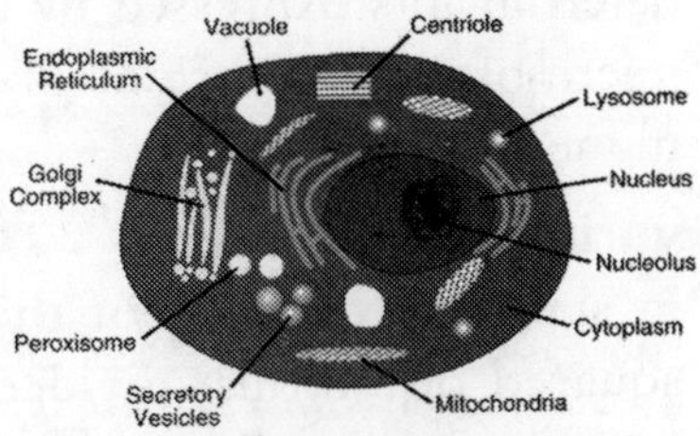

■ **vacuum**

vacuum is created through use of a vacuum pump or aspirator pump, to facilitate specific biological preparations, such as inclusions or disinfection of material for *in vitro* culture, etc.

■ **variable domains**

regions of antibody chains that have different amino acid sequences in different antibody molecules. These regions are responsible for th antigen-binding specificity of- the antibody molecule.

■ **variable expressivity**

variation in the phenotype caused by different alleles of the same gene and/or by the action of other genes and/or by the action of non-genetic factors. *cf* expressivity.

variable surface glycoprotein (VSG)
one of a battery of antigenic determinants expressed by a micro-organism to elude immune detection.

variance
in statistics: the sum of the squared deviations, divided by one less than the number of observations. A statistical measure of variation in a population.

variant
an organism that is genetically different from the wild type organism. a.k.a. mutant.

variation
differences between individuals within a population or among populations.

variegated
plants having both green and albino tissues. This difference in colour may result from viral infection, nutritional deficiency, or may be under genetic or physiological control.

variety
a naturally occurring subdivision of a species, with distinct morphological characters and given a Latin name according to the rules of the International Code of Nomenclature. A taxonomic variety is known by the first validly published name applied to it so that nomenclature tends to be stable. *cf* cultivar; pathovar.

vascular
referring to any plant tissue or region consisting of or giving rise to conducting tissue, e.g., bundle, cambium, ray.

vascular bundle
also known as fascicle .A strand of tissue containing primary xylem and primary phloem (and procambium if present) and frequently enclosed by a bundle sheath of parenchyma or fibres.

vascular cambium
in biennials and perennials, cambium giving rise to secondary phloem and secondary xylem.

vascular plants
plants possessing organised vascular tissues.

vascular system
1. a specialised network of vessels for the circulation of fluids throughout the body tissue of an animal.

2. the system of vascular tissue in plants.

■ **vascular tissue**

the tissue that conducts water and nutrients throughout the plant body in higher plants.

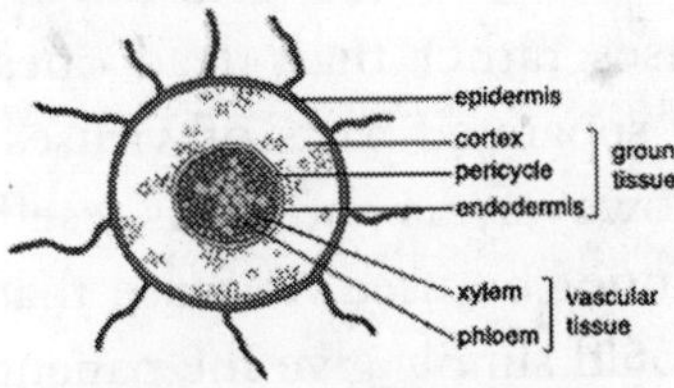

Figure 3. Cross section of a root

■ **vector**

1. an organism, usually an insect, that carries and transmits disease-causing organisms.
2. a plasmid or phage that is used to deliver selected foreign DNA for cloning and in gene transfer.

■ **vegetative propagation**

same as asexual or non-sexual propagation.

■ **vehicle**

the host organism used for the replication or expression of a cloned gene or other sequence The term is little used and is often confused with vector. *See* vector.

■ **velocity density gradient centrifugation**

a procedure used to separate macromolecules based on their rate of movement through a density gradient.

■ **velogenetics**

the combined use of marker-assisted selection (q.v.) and embryo technologies such as OPU (q.v.), IVM (q.v.) and IVF (q.v.), in order to increase the rate of genetic improvement in animal populations.

■ **vermiculite**

coarse aggregate material made from expanded mica having a high cation exchange capacity and high water-holding capacity and used as a rooting medium and a soil additive.

■ **vernalisation**

the process by which floral induction in some plants is promoted by exposing the plants to chilling for a certain duration.

■ **vessel**

1. a series of xylem elements whose function is to conduct water and nutrients in plants.

2. a container, such as a Petri dish or test tube, used for tissue culture.

■ **vessel element**

a type of cell occurring within the xylem of flowering plants. Many are water-conducting vessels.

■ **viability**

the capability to live and develop normally.

■ **viability test**

assay of the number or percent of living cells or plants in a population that have been given a specific treatment; such as ascertaining cell viability after cryopreservation treatment, or for seeds.

■ **viable**

capable of germinating, living, growing and developing.

■ **vibrio**

comma-shaped bacterium.

■ **vir genes**

a set of genes on a Ti plasmid that prepare the T-DNA segment for transfer into a plant cell.

■ **viral coat protein**

protein present in the outer layer of a virus.

■ **viral oncogene**

a gene in the viral genome that contributes to malignancies in vertebrate hosts..

■ **viral pathogen**

a disease-causing virus.

■ **viral vaccines**

vaccines consisting of live viruses rather than dead ones or separated parts of viruses. However, as the virus itself cannot be used, because that would simply give the patient the disease, the virus is genetically engineered so that it elicits the immune response to the viral pathogen without causing the disease itself.

two genetic engineering methods can be used. The first is to make the disease virus harmless, but still able to replicate in cultured animal cells. This is similar to producing an 'attenuated' virus, i.e., one which has been grown in the laboratory until it loses its ability to cause disease. However, the genetic engineering route seeks to make sure that the attenuated virus has no chance of mutating back to a wild type, pathogenic virus, by deleting whole genes or replacing key regions of genes with completely different genetic material.

the second approach is to clone the gene for a protein from the pathogenic virus into another, harmless virus, so that the result 'looks' like the pathogenic virus but does not cause disease.

■ **virion**

an infectious virus particle. A plant pathogen that consists of a naked RNA molecule of approximately 250-350 nucleotides, whose extensive base pairing results in a nearly correct double helix.

■ **viroid**

an infectious entity similar to a virus but smaller, consisting only of a strand of nucleic acid without the protein coat characteristic of a virus.

■ **virulence**

the degree of ability of an organism to cause disease. The relative infectiousness of a bacterium or virus, or its ability to overcome the resistance of the host metabolism.

■ **virulent phage**

a phage (virus) that destroys the host (bacterium).

■ **virus**

a poisonous or slimy liquid .An infectious particle composed of a protein capsule and a nucleic acid core (DNA or RNA), which is dependent on a host organism for replication. The DNA or a double-stranded DNA copy of an RNA virus genome is integrated into the host chromosome during lysogenic infectio or replicated during the ystic cycle.

■ **virus-free**

plant, animal, cell, tissue or meristem which exhibits no viral symptoms or contains no identifiable virus-particles.

■ **virus-tested**

description of a organism or a cell stock certified as being free of certain specified viruses following recognised procedures of virus diagnosis.

■ **visible light**

the part of the electromagnetic spectrum with wavelengths between 380 nm and

750 nm and perceived by the human eye.

vitamin B complex

a large group of water soluble vitamins that function as co-enzymes, including thiamine (B_1); riboflavin or vitamin G (B_2); niacin or nicotinic acid (B_3); pantothenic acid (B_5); pyridoxine (B_6); cyanocobalamin (B_{12}); biotin or vitamin H; folic acid or vitamin M (Bc); inositol; choline; and others.

vitamins

naturally occurring organic substances required by living organisms in relatively small amounts to maintain normal health, and which are added to tissue culture media to enhance growth, usually acting as enzyme co-factors.

vitrified

cultured tissue having leaves and sometimes stems with a glassy, transparent or wet and often swollen appearance. The process of vitrification is a general term for a variety of physiological disorders that lead to shoot tip and leaf necrosis.

vivipary

1. a form of reproduction in animals in which the developing embryo obtains its nourishment directly from the mother via a placenta or by other means.

2. a form of asexual reproduction in certain plants, in which the flower develops into a budlike structure that forms a new plant when detached from the parent.

3. the development of young plants in the inflorescence of the parent plant.

Vmax

the maximal rate of an enzyme-catalysed reaction. V_{max} is the product of E_o (the total amount of enzyme) times the value of K_{eat} (the catalytic rate constant).

VNTR

variable number tandem repeat. A short DNA sequence that is present as tandem repeats and in highly variable copy number.

volatilisation

the conversion of a solid or liquid into a gas or vapour.

■ **w/v**

weight per volume; the weight of a constituent in 100 cm^3 of solution, expressed as a percentage.

■ **walking**

a method for cloning large regions of a chromosome. Starting from a known site, a gene library is screened for clones that hybridise to DNA probes taken from the ends of the first clone. These clones are then isolated, and their ends used to screen the library again. These clones are then isolated and their ends used, and so on. *See* gene walking; chromosome walking.

■ **wall pressure**

pressure that a cell wall exerts against the turgor of the cell contents. Wall pressure is equal and opposite to the turgor potential.

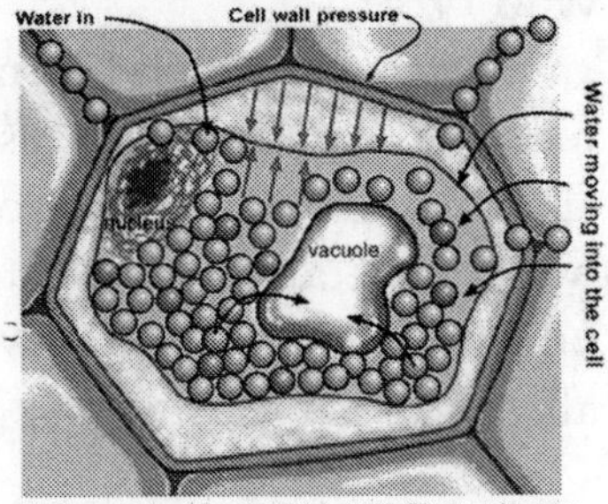

Cell becomes turgid - gets rigid due to it's filling up with water.

■ **wash-out**

the loss of the slower growing micro-organism when two organisms are being grown together.

■ **water potential**

refers to the difference between the activity of water molecules in pure distilled water at atmospheric pressure and 30°C (standard conditions), and the activity of water molecules in any other system. The activity of these water molecules may be greater (positive) or less (negative) than the activity of the water molecules under standard conditions.

■ **water stress**

the condition when plants are unable to absorb enough wate- r to replace that lost by tra spiration. The results may b wilting, cessation of growt , or even death of the pla t or plant parts.

■ **wax**

esters of alcohol higher than glycerol, which are insoluble in water and difficult to hydrolyze; wax forms protective waterproof layers on leaves, stems, fruits, animal fur and integuments of insects.

■ **weed**

simply any plant growing where it is not wanted. In agriculture, used for a plant which has good colonizing capability in a disturbed environment, and can usually compete with a cultivated species therein. Weeds are typically considered as unwanted, economically useless or pest species.

■ **weediness**

in agriculture, the ability of a plant to colonise a disturbed habitat and compete with cultivated species.

■ **western blot**

a technique in which protein is transferred from an electrophoretic gel to a cellulose or nylon support membrane following electrophoresis. A particular protein molecule can then be identified by probing the blot with a radio labelled antibody which binds only the specific protein to which the antibody was prepared. Usefu , for example, for measurem nt of levels of production of a specific protein in a particular tissue or at particular developmental stage. *cf* blot; northern blot; Southern blot.

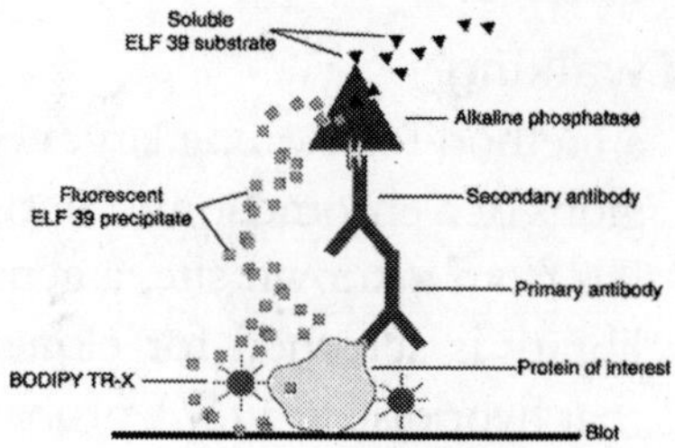

■ **wet weight**

the gross weight of a product with its full water content, or weight of fully hydrated tissue. *See* fresh weight.

■ **wetting agent**

a substance that improves surface contact by reducing the surface tension of a liquid: e.g., Triton X-10TM added to disinfecting solutions promotes the disinfestation process. *See* detergent; surfactant.

■ **wild type**

an organism as found in nature; the dominant allele usually found in nature, and from which mutations produce other dominants or recessives alleles.

■ wilt

drooping of stems and foliage due to loss of water and decreased turgidity of cells. May be caused by water stress or by disease.

■ wilting point

the moisture content of soil at which plants start to wilt, but not to the extent that they fail to recover when placed in a humid atmosphere. *cf.* permanent wilting point.

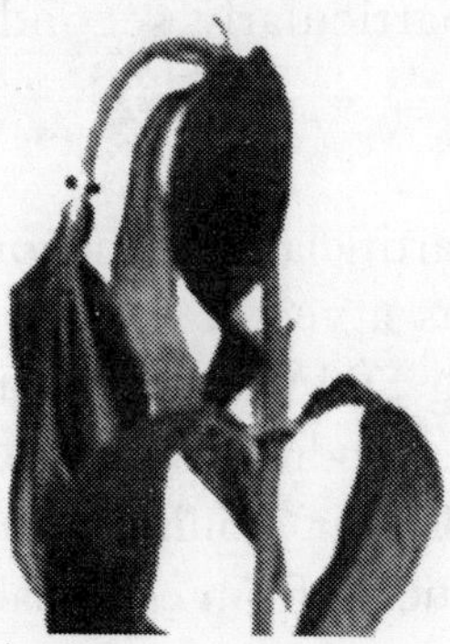

wobble hypothesis

an explanation of how one tRNA may recognise more than one codon. The first two bases of the mRNA codon and anticodon pair properly, but the third base in the anticodon has some wobble that permits it to pair with more than one base.

■ x

the basic number of chromosomes in a polyploid series, monoploid = x; diploid = 2x; triploid = 3x; etc.

■ xanthophyll

a yellow chloroplast pigment.

■ X-chromosome

a chromosome associated with sex determination. In most animals, the female has two, and the male has one X chromosome. The opposite occurs in birds, in which the equivalent is the Z chromosome.

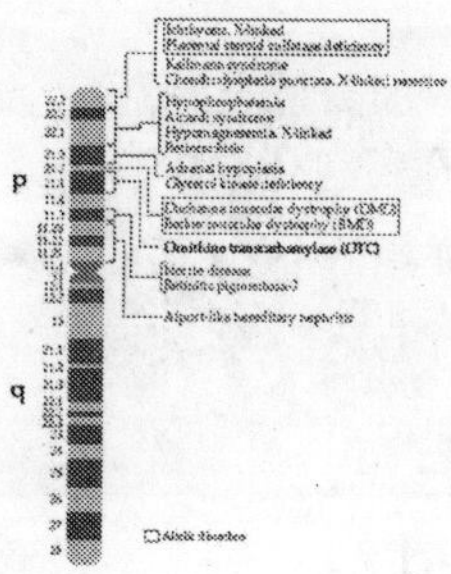

■ xenia

the immediate effect of pollen on some characters of the endosperm.

■ xenobiotic

a chemical compound that is not produced by living organisms; a manufactured chemical compound.

xenotransplantation
the transplantation of tissue from one species to another species, typically from non-human mammals to humans. This technology has become very important because of a worldwide shortage of human organs for humans requiring a new organ. The most popular non-human species involved in this technology is the pig.

xerophyte
a plant very resistant to drought or that lives in very dry places.

X-linked
the presence of a gene on the X chromosome. a.k.a. X-linkage.

X-linked disease
a genetic disease caused by an allele at a locus on the X-chromosome. In X-linked recessive conditions, a normal female "carrier" passes on the allele on her X chromosome to an affected son.

X-ray crystallography
the deduction of crystal structure from analysis of the diffraction pattern of X-rays passing through a pure crystal of a substance.

xylem
a complex tissue specialised for efficient conduction of water and mineral nutrients in solution. Xylem may also function as a supporting tissue, particularly secondary xylem.

YAC
yeast artificial chromosome, used as a vector system for cloning DNA fragments that can be hundreds of kilobases long. Linear cloning vectors constructed from essential elements of yeast chromosomes. They can accommodate foreign DNA inserts of 200 to 500 kb.

Y-chromosome
the partner of the X-chromosome in the male of many animal species.

yeast
a unicellular ascomycete fungus, commonly found as a contaminant in plant tissue culture.

yeast cloning vectors
the yeasts, and especially *Saccaromyces cerevisiae*, are favourite organisms in which to clone and express DNA. They are eukaryotes, and so can splice out introns, the non-coding sequences in the middle of many eukaryotic genes.

yeast episomal vector
a cloning vector for the yeast *Saccaromyces cerevisiae* that uses 2mm plasmid as origin of replication and is maintained as an extrachromosomal nuclear DNA molecule.

yeast extract
a mixture of substances from yeast. *cf* organic complex; undefined.

Z-DNA
also known as zig-zag DNA. A form of DNA duplex in which the double helix is wound in a left-hand, instead of a right-hand, manner. DNA adopts the Z configuration when purines and pyrimidines alternate on a single strand, e.g., 5´CGCGCGCG3´ or 5´CACACACACA3´ 3´GCGCGCGC5´ 3´GTGTGTGTGT5´

zone of elongation
the section of the young root or shoot just behind the apical meristem, in which the cells are enlarging and elongating rapidly.

zoo blot
hybridisation of cloned DNA from one species to DNA from other organisms to determine the extent to which the cloned DNA is evolutionarily conserved.

zoo FISH
fluorescent *in situ* hybridisation of DNA from one species on metaphase chromosomes of another species. Typically, the hybridisation is done separately for DNA libraries representing each chromosome. The result is a fascinating picture of the regions of chromosomal homology between species.

■ **zoospore**

a spore that possesses flagella and is therefore motile.

■ **zygonema**

stage in meiosis during which synapsis occurs; coming after the leptotene stage and before the pachytene stage in the meiotic prophase.

■ **zygospore**

a thick-walled resistant spore developing from a zygote resulting from the fusion of isogametes.

■ **zygote**

a diploid cell formed by the fusion of two haploid gametes during fertilisation in eukaryotic organisms with sexual reproduction. It is the first cell of the new individual.

■ **zymogen**

inactive enzyme precursor that after secretion is chemically altered to the active form of the enzyme.

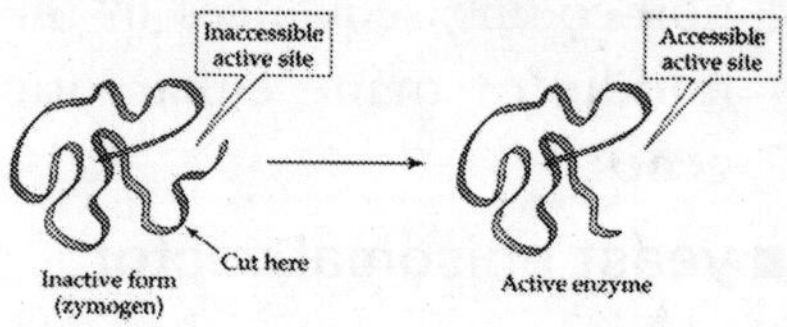

Other Books on

LOTUS ILLUSTRATED DICTIONARIES

1. Dictionary of New Words (New)	125/-
2. Dictionary of Veterinary Sciences (New)	125/-
3. Dictionary of Synonyms and Antonyms (New)	150/-
4. Dictionary of Idioms and Phrases (New)	150/-
5. First English Dictionary	125/-
6. Astronomy	125/-
7. Agriculture	125/-
8. Anthropology	125/-
9. Archaeology	125/-
10. Architecture	125/-
11. Art	125/-
12. Banking Finance & Accounting	125/-
13. Bio-Chemistry	125/-
14. Business Administration	125/-
15. Bio-Technology	125/-
16. Biology	125/-
17. Botany	125/-
18. Culture	125/-
19. Chemical Engg.	125/-
20. Chemistry	125/-
21. Civil Engineering	125/-
22. Computer Science	125/-
23. Commerce	125/-
24. Cooking & Food	125/-
25. Ecology	125/-
26. Economics	125/-
27. Education	125/-
28. Electrical Engineering	125/-
29. Electronic & Telecommunication	125/-
30. Environmental Studies	125/-
31. Festival	125/-
32. Geography	125/-

NOTES

NOTES